Documentation of the Seawater Intrusion (SWI2) Package for MODFLOW

By Mark Bakker, Frans Schaars, Joseph D. Hughes, Christian D. Langevin, and Alyssa M. Dausman

Chapter 46 of
Section A, Groundwater
Book 6, Modeling Techniques

Groundwater Resources Program

Techniques and Methods 6–A46

U.S. Department of the Interior
U.S. Geological Survey

U.S. Department of the Interior
SALLY JEWELL, Secretary

U.S. Geological Survey
Suzette M. Kimball, Acting Director

U.S. Geological Survey, Reston, Virginia: 2013

For more information on the USGS—the Federal source for science about the Earth, its natural and living resources, natural hazards, and the environment, visit http://www.usgs.gov or call 1–888–ASK–USGS.

For an overview of USGS information products, including maps, imagery, and publications, visit http://www.usgs.gov/pubprod

To order this and other USGS information products, visit http://store.usgs.gov

Suggested citation:
Bakker, Mark, Schaars, Frans, Hughes, J.D., Langevin, C.D., and Dausman, A.M., 2013, Documentation of the seawater intrusion (SWI2) package for MODFLOW: U.S. Geological Survey Techniques and Methods, book 6, chap. A46, 47 p., http://pubs.usgs.gov/tm/6a46/.

Preface

The Seawater Intrusion (SWI2) Package was written for use with the U.S. Geological Survey (USGS) MODFLOW-2005 groundwater model. The SWI2 Package is designed to simulate regional seawater intrusion in coastal aquifer systems. The performance of this computer program has been tested in models of hypothetical and actual coastal aquifers; however, future applications of the programs may reveal errors that were not detected in the test simulations. Users are requested to notify the USGS if errors are found in the documentation report or in the computer program.

Although the computer program has been written and used by the USGS, no warranty, expressed or implied, is made by the USGS or the United States Government as to the accuracy and functionality of the program and related program material. Nor shall the fact of distribution constitute any such warranty, and no responsibility is assumed by the USGS in connection therewith.

MODFLOW-2005, the SWI2 Package, and other groundwater programs are available online from the USGS at the following address:

http://water.usgs.gov/software/lists/groundwater/

Acknowledgments

The authors thank Theo Olsthoorn of Waternet and the Delft University of Technology for his suggestions and support. Vaughan Voller of the University of Minnesota provided valuable suggestions for the tip and toe tracking algorithm. Debbie Borden of the University of Georgia extensively tested the tip and toe tracking algorithm. The authors are also indebted to Delywn Oki and Kolja Rotzoll for providing several test problems that significantly improved the SWI2 Package.

The authors are also grateful to the technical reviewers of this report, including Hedeff I. Essaid and Kolja Rotzoll of the U.S. Geological Survey.

Contents

Figures

Conversion Factors and Abbreviations

SI to Inch/Pound

Multiply	By	To obtain
Length		
meter (m)	3.281	foot (ft)
Volume		
cubic meter (m³)	35.31	cubic foot (ft³)
Flow rate		
cubic meter per day (m³/d)	35.31	cubic foot per day (ft³/d)
Density		
kilogram per cubic meter (kg/m³)	0.06242	pound per cubic foot (lb/ft³)
Areal recharge		
millimeter per year (mm/yr)	0.003937	inch per year (in/yr)
millimeter per day (mm/d)	0.003937	inch per day (in/d)
Hydraulic conductivity		
meter per day (m/d)	3.281	foot per day (ft/d)
Hydraulic gradient		
meter per kilometer (m/km)	5.27983	foot per mile (ft/mi)
Transmissivity*		
meter squared per day (m²/d)	10.76	foot squared per day (ft²/d)
Leakance		
meter per day per meter [(m/d)/m]	1	foot per day per foot [(ft/d)/ft]

*Transmissivity: The standard unit for transmissivity is cubic foot per day per square foot times foot of aquifer thickness [(ft3/d)/ft2]ft. In this report, the mathematically reduced form, foot squared per day (ft2/d), is used for convenience. Vertical coordinate information is referenced to the National Geodetic Vertical Datum of 1929 (NGVD 29).

Horizontal coordinate information is referenced to the North American Datum of 1983 (NAD 83)

Elevation, as used in this report, refers to the distance above a vertical datum.

Abbreviations

ASCII   American Standard Code for Information Exchange

GHB   general head boundary

SWI   Seawater Intrusion (Package)

TVD   total variation diminishing

Documentation of the Seawater Intrusion (SWI2) Package for MODFLOW

By Mark Bakker,[1] Frans Schaars,[2] Joseph D. Hughes,[3] Christian D. Langevin,[3] and Alyssa M. Dausman[3]

Abstract

The SWI2 Package is the latest release of the Seawater Intrusion (SWI) Package for MODFLOW. The SWI2 Package allows three-dimensional vertically integrated variable-density groundwater flow and seawater intrusion in coastal multiaquifer systems to be simulated using MODFLOW-2005. Vertically integrated variable-density groundwater flow is based on the Dupuit approximation in which an aquifer is vertically discretized into zones of differing densities, separated from each other by defined surfaces representing interfaces or density isosurfaces. The numerical approach used in the SWI2 Package does not account for diffusion and dispersion and should not be used where these processes are important. The resulting differential equations are equivalent in form to the groundwater flow equation for uniform-density flow. The approach implemented in the SWI2 Package allows density effects to be incorporated into MODFLOW-2005 through the addition of pseudo-source terms to the groundwater flow equation without the need to solve a separate advective-dispersive transport equation. Vertical and horizontal movement of defined density surfaces is calculated separately using a combination of fluxes calculated through solution of the groundwater flow equation and a simple tip and toe tracking algorithm.

Use of the SWI2 Package in MODFLOW-2005 only requires the addition of a single additional input file and modification of boundary heads to freshwater heads referenced to the top of the aquifer. Fluid density within model layers can be represented using zones of constant density (stratified flow) or continuously varying density (piecewise linear in the vertical direction) in the SWI2 Package. The main advantage of using the SWI2 Package instead of variable-density groundwater flow and dispersive solute transport codes, such as SEAWAT and SUTRA, is that fewer model cells are required for simulations using the SWI2 Package because every aquifer can be represented by a single layer of cells. This reduction in number of required model cells and the elimination of the need to solve the advective-dispersive transport equation results in substantial model run-time savings, which can be large for regional aquifers. The accuracy and use of the SWI2 Package is demonstrated through comparison with existing exact solutions and numerical solutions with SEAWAT. Results for an unconfined aquifer are also presented to demonstrate application of the SWI2 Package to a large-scale regional problem.

Introduction

The SWI2 Package is designed to simulate regional seawater intrusion in coastal aquifer systems by representing variable-density flow with discrete zones of uniform or linearly varying density. A number of computer codes exist for the simulation of seawater intrusion in coastal aquifers. In general, variable-density seawater intrusion models that track saltwater movement may be divided into interface and dispersive solute transport models. In interface models, freshwater and seawater are separated by an interface (for example, Wilson and Sa da Costa, 1982; Essaid, 1990; Taylor and Person, 1998). In dispersive solute transport models, fluid density can vary continuously or from cell to cell in a model domain. Some programs solve the flow and transport equations simultaneously, such as SUTRA (Voss and Provost, 2010) and FEFLOW (Diersch and Kolditz, 2002). Others solve the transport equations separately using particle-based or finite-difference methods and compute a new flow field as necessary to represent a changing density field; examples of such programs include MOCDENS3D (Oude Essink, 2001) and SEAWAT (Langevin and others, 2008).

The original SWI Package (Bakker and Schaars, 2011) was developed for MODFLOW-2000 (Harbaugh and others, 2000) and is based on the vertically integrated variable-density formulation for groundwater flow by Bakker (2003). Vertically integrated fluxes were used for variable-density flow prior to the work of Bakker (2003). Weiss (1982) and Maas and Emke (1988) used vertically integrated fluxes to describe instantaneous variable-density flow fields. Strack (1995) used vertically integrated fluxes to develop a potential flow formulation for variable-density flow. The SWI2 Package incorporates several improvements to the original SWI Package, including the ability to simulate vertical interface movement through

[1]Delft University of Technology, Delft, The Netherlands.

[2]Artesia Water Research, Schoonhoven, The Netherlands.

[3]U.S. Geological Survey.

multiple model layers, adaptive time stepping, and complete reporting of the mass balance for each density zone.

The main advantage of the formulation used in the SWI2 Package compared to models that solve flow and transport equations is that the SWI2 Package represents three-dimensional vertically integrated variable-density flow without the need to discretize the aquifer vertically. Instead, the Dupuit approximation is adopted and each aquifer (represented as a single model layer) is discretized vertically into zones having different densities. As a result, numerical simulations using the SWI2 Package require far fewer cells than dispersive solute transport simulations. Adoption of the Dupuit approximation is interpreted to mean that within an aquifer the resistance to vertical flow is neglected and there is no vertical head gradient (hydrostatic conditions).

Another benefit of the formulation used in the SWI2 Package is that it does not require modifications to most of the existing routines in MODFLOW-2005 (Harbaugh, 2005). The SWI2 Package is designed to be independent of the other MODFLOW-2005 packages; the effects of variable-density flow are added to the system of equations as SWI2 pseudo-source terms. After the groundwater flow equation is solved, a separate solution is required to simulate horizontal and vertical movement of surfaces separating zones of different densities. Solution of the density surfaces is done with the existing MODFLOW-2005 solvers.

Because SWI2 makes it possible to simulate vertically integrated variable-density groundwater flow using one model layer per aquifer, seawater intrusion can be simulated in existing regional MODFLOW-2005 models with limited modifications. The necessary modifications to existing MODFLOW-2005 models include the addition of a single SWI2 Package input file and modification of boundary conditions and constant heads representing coastal boundaries to freshwater heads at the top of the aquifer. For most coastal problems, seawater boundaries will be referenced to a sea level at or close to zero. This may not be the case, however, if model elevations are referenced to an arbitrary datum not based on sea level (example simulation 6 uses an arbitrary datum). The generalized approach used in the SWI2 Package is applicable, as part of MODFLOW-2005, to a wide range of coastal settings.

The SWI2 Package is a powerful MODFLOW-2005 addition that permits the simulation of regional variable-density groundwater flow. Because this report describes only features of the SWI2 Package and necessary modifications to other MODFLOW-2005 packages, readers are encouraged to use this report to supplement existing documentation of MODFLOW-2005 (Harbaugh, 2005).

Currently (2013), MODFLOW-2005 (Harbaugh, 2005) is the most recent version of the MODFLOW code. MODFLOW was originally developed in the 1980s (McDonald and Harbaugh, 1988) and has been continuously updated (Harbaugh and McDonald, 1996; Harbaugh and others, 2000). Unless otherwise noted herein, the term "MODFLOW" refers to the MODFLOW-2005 version of the code.

Development of the SWI2 Package was funded by the U.S. Geological Survey Groundwater Resources Program. Development of the original SWI Package was made possible through grants of the Georgia Coastal Incentive Grants Program, administered by the Georgia Department of Natural Resources, and through financial support of the Amsterdam Water Supply (now Waternet) in The Netherlands.

Purpose and Scope

This report serves as documentation for the Seawater Intrusion (SWI2) Package. Use of the SWI2 Package and the modifications required to simulate vertically integrated variable-density groundwater flow in existing MODFLOW models are described. Instructions for running MODFLOW with the SWI2 Package and the format for input datasets are provided (appendix 1). Finally, benchmark and demonstration problems are described, and results from MODFLOW with the SWI2 Package are presented. Seven example problems are presented and include evaluation of (1) a rotating freshwater-seawater interface, (2) a rotating brackish zone, (3) saltwater intrusion in a two-aquifer coastal aquifer system, (4) upconing in a two-aquifer island system, (5) upconing in a radial flow system, (6) the effect that SWI2 assumptions have on simulated saltwater intrusion in a two-aquifer coastal aquifer system, and (7) saltwater intrusion in a regional model of the shallow, unconfined aquifer underlying Cape Cod, Massachusetts.

Mathematical Formulation of Vertically Integrated Variable-Density Groundwater Flow in Coastal Aquifers

The governing equations developed for the SWI2 Package are used to represent vertically integrated variable-density groundwater flow in MODFLOW. The formulation presented here is only appropriate for water in the liquid phase having densities less than or equal to seawater. The SWI2 Package is not intended for the simulation of the combined flow of freshwater and brines, because variations in viscosity are neglected. After a mathematical derivation of the equations implemented in the SWI2 Package, an alternative derivation is presented to facilitate understanding of the mathematical formulation.

Conceptual Model

A schematic vertical cross section of saltwater intruding a coastal aquifer is shown in figure 1A. Freshwater flows from right to left, towards the sea, and is separated from the underlying saltwater by an interface. The coastal aquifer is discretized into one model layer of 14 cells (fig. 1B); the 10 cells on the right side of the model contain both freshwater and saltwater, whereas the 4 cells on the left side only contain saltwater. Consider a pipe open to the bottom of the ocean (fig. 1*A* inset). When the pipe is filled with saltwater, the water level in the pipe will be equal to sea level. When the pipe is filled with freshwater, the water level in the pipe will be higher than sea level (because freshwater is less dense than saltwater). This type of head is referred to as freshwater head.

In coastal aquifers, groundwater density is a function of salinity. In the SWI2 Package, the density in each aquifer is divided vertically into a number of discrete zones bounded by three-dimensional surfaces. A schematic vertical cross section of this conceptualization is shown in figure 2*A*; the thick dashed lines represent the surfaces separating the zones. The elevation of each surface is a unique function of the horizontal

coordinates. The SWI2 Package has two options. For the first option, the water has a constant density in each zone, the density is discontinuous from zone to zone, and each surface represents an interface (fig. 2*B*). Multiple interfaces may be used to separate, for example, freshwater from brackish water, and brackish water from saltwater; this option is referred to herein as the "stratified" option. For the second option, water density varies linearly in the vertical direction in each zone, density is continuous from zone to zone, and the surfaces bounding the zones are density isosurfaces (fig. 2*C*); water density does not vary in the freshwater and saltwater zones (fig. 2*D*). This option is referred to herein as the "continuous" option.

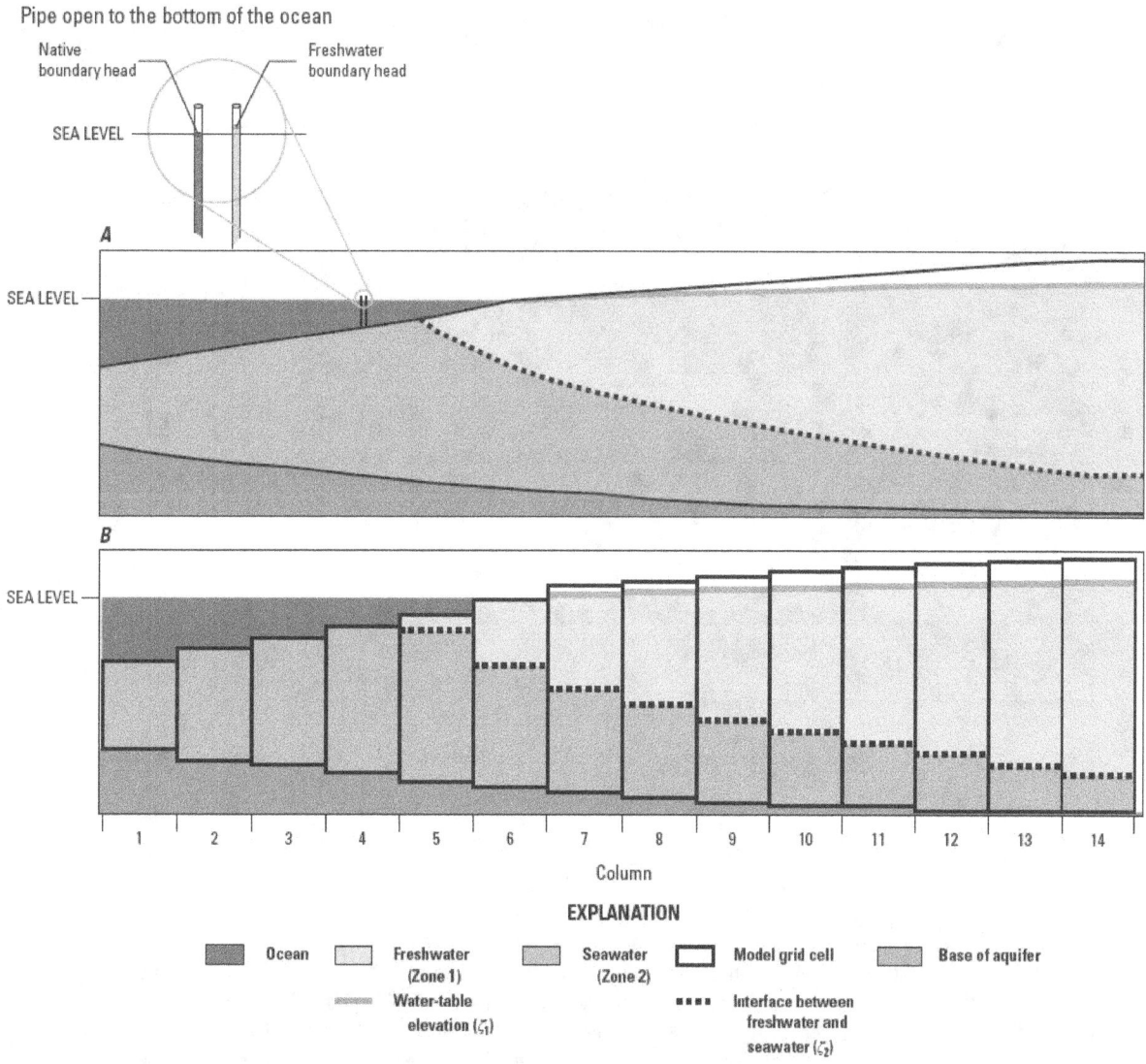

Figure 1.   *A*, cross section through a coastal aquifer with freshwater flowing from right to left and discharging seaward of the coastline; *B*, cross section through a coastal aquifer discretized using a single model layer. The position of the freshwater-seawater interface is also shown on *A* and *B*.

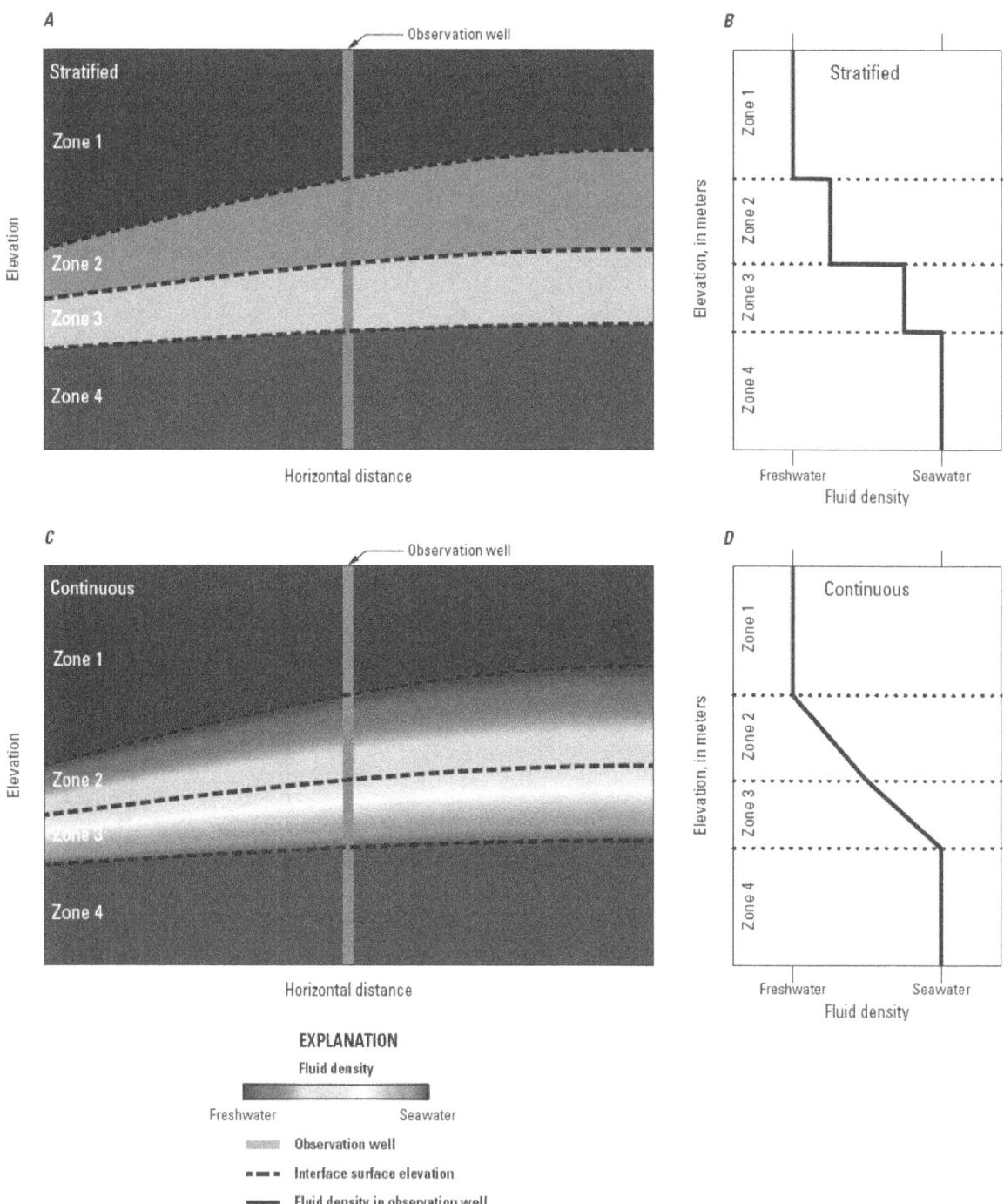

Figure 2.   Aquifer vertical fluid density distribution options available in the SWI2 Package: *A,* Vertical cross section of an aquifer using the discontinuous, stratified density (``stratified'') option; *B,* vertical fluid density distribution in the observation well for an aquifer discretized using the discontinuous, stratified density (``stratified'') option; *C,* vertical cross section of an aquifer using the continuous piece-wise linear density (``continuous'') option; *D,* vertical fluid density distribution in the observation well for an aquifer discretized using the continuous piece-wise linear density (``continuous'') option.

Governing Equations for Variable-Density Groundwater Flow

A Cartesian x, y, z coordinate system is adopted with the z-axis pointing vertically upward. Darcy's law for variable-density flow may be written as (Post and others, 2007)

$$q_x = -K\frac{\partial h_f}{\partial x} \qquad q_y = -K\frac{\partial h_f}{\partial y} \qquad q_z = -K\left(\frac{\partial h_f}{\partial z} + v\right), \quad (1)$$

where

q_x, q_y, q_z	are the components of the three-dimensional specific discharge vector [L/T],
K	is the freshwater hydraulic conductivity [L/T],
h_f	is the freshwater head [L], and
v	is the dimensionless density [unitless], defined as

$$v = \frac{\rho - \rho_f}{\rho_f}, \quad (2)$$

where

ρ	is the fluid density [M/L³], and
ρ_f	is the density of freshwater [M/L³].

The derivation is presented here for an isotropic aquifer, but may be rewritten for an anisotropic aquifer where the principal components of the hydraulic conductivity tensor point in the x, y, and z directions, as required by MODFLOW.

The freshwater head is a function of the three spatial coordinates and is defined as

$$h_f(x, y, z) = \frac{p(x, y, z)}{\rho_f g} + z, \quad (3)$$

where

p	is the fluid pressure [M/L·T²],
g	is the acceleration of gravity [L/T²].

The freshwater hydraulic conductivity is defined as

$$K = \frac{k\rho_f g}{\mu_f}, \quad (4)$$

where

k	is the intrinsic permeability of the aquifer [L²], and
μ_f	is the viscosity of freshwater [M/L·T].

Variations in viscosity are neglected because the viscosity of freshwater is approximately equal to that of seawater. The difference between the hydraulic conductivity of freshwater and seawater is generally only a few percent and is therefore neglected.

Basic Approximations

Four basic approximations are made when mathematically deriving the formulations used in the SWI2 Package:

1. The Dupuit approximation is adopted and is interpreted to mean that the resistance to flow in the vertical direction within a single aquifer is neglected (for example, Strack, 1989, p.36).

2. The mass balance equation is replaced with a volume balance equation in the computation of the flow field (the Boussinesq-Oberbeck approximation), and density effects are taken into account only through Darcy's law. Holzbecher (1998, p. 32) provides a complete derivation of the Boussinesq-Oberbeck approximation.

3. The effects of dispersion and diffusion are considered negligible.

4. Density inversions are allowed between aquifers but not within an aquifer. A density inversion occurs when saltier, denser water overlies fresher, less-dense water and often results in the vertical growth of fingers (for example, Wooding, 1969). The SWI2 Package is designed for modeling regional seawater intrusion, which generally exceeds the scale of fingers that develop as a result of density inversions.

Mathematical Derivation of Vertically Integrated Variable-Density Groundwater Flow in Aquifers

The derivation of vertically integrated groundwater flow in aquifers presented in this section is based on Bakker (2003); an alternative derivation is presented later. Groundwater in an aquifer is discretized vertically into $N+1$ surfaces that bound N zones having unique densities. Zones and surfaces are numbered from the top down and specific surfaces and zones are referred to using a lowercase n; zone n is bounded on top by surface n and below by surface $n+1$ (fig. 3); thus, zone 1 is bounded on top by surface 1 and on the bottom by surface 2. The elevation of surface n is represented by the function $\zeta_n(x,y)$. The elevation of the top of the saturated aquifer is defined as ζ_1; this can be either the elevation of the top of the aquifer (layer) if the aquifer is confined or the elevation of the water table. The bottom elevation of the aquifer is defined as ζ_{N+1}.

Flow in an aquifer is expressed in terms of the freshwater head $h(x, y)$ at the saturated top of an aquifer and in the elevations of the surfaces between the zones (ζ_2 through ζ_N). The Dupuit approximation is adopted, which means that the vertical pressure distribution is approximated as hydrostatic, and the freshwater head, h_f, at any elevation z may be expressed in terms of the freshwater head $h(x, y)$ at either the top of the aquifer or the top of the water surface as

$$h_f(x, y, z) = h(x, y) + \int_z^{\zeta_1} v(x, y, z')\, dz', \quad (5)$$

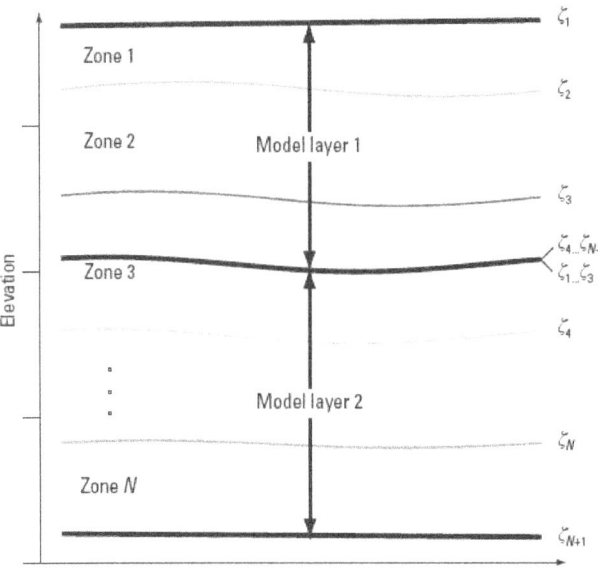

Figure 3. Two aquifers discretized into two model layers showing the numbering of the zones and surfaces in each model layer.

where z' is the vertical coordinate (or the dummy variable of integration) [L]. Henceforth, h refers to the freshwater head at the top of the aquifer or the top of the water surface. The horizontal specific discharge vector $\vec{q} = \left(q_x, q_y\right)$ [L/T], defined as a function of the vertical coordinate z in an aquifer, is obtained by substituting equation 5 into equation 1:

$$\vec{q} = -K\nabla h - K\nabla \int_z^{\zeta_1} v\left(x,y,z'\right)dz',\qquad(6)$$

where ∇ is the two-dimensional gradient vector $\nabla = \left(\partial/\partial x, \partial/\partial y\right)$. The vertically integrated horizontal flow vector for all zones below surface n is called $\vec{U}_n$ and is obtained from integrating the horizontal specific discharge vector from the bottom of the aquifer (ζ_{N+1}) to surface n (ζ_n).

$$\vec{U}_n = \int_{\zeta_{N+1}}^{\zeta_n} \vec{q}\,dz.\qquad(7)$$

Because surface 1 is the saturated top of the aquifer, $\vec{U}_1$ represents the total horizontal flow through an area having a unit width and a height equal to the thickness of the aquifer [L²/T]; the continuity of flow in the aquifer may therefore be written as

$$\nabla \vec{U}_1 = -S\frac{\partial h}{\partial t} - q_{z,t} + q_{z,b} + \gamma,\qquad(8)$$

where

S	is the storage coefficient of the aquifer [unitless],
$q_{z,t}, q_{z,b}$	are the vertical specific discharge [L/T] at the top and bottom of the aquifer, respectively, and
γ	is a source term [L/T].

Similarly, continuity of flow below surface n may be written as

$$\nabla \vec{U}_n = -n_e \frac{\partial \zeta_n}{\partial t} + \gamma_n \qquad n = 2,3,\dots,N,\qquad(9)$$

where

n_e	is the effective porosity [unitless] and
γ_n	represents all source terms below surface n [L/T] and may include leakage from underlying or overlying aquifers.

Combining equations 5 to 9 yields the following set of N partial differential equations. The continuity of flow equation below surface 1 is

$$\nabla\left(T\nabla h\right) = S\frac{\partial h}{\partial t} + q_{z,t} - q_{z,b} - \gamma + R,\qquad(10)$$

where

T	is the transmissivity of the aquifer [L²/T], and
R	is the pseudo-source term [L/T] below surface 1.

The pseudo-source term, R, is the result of the flux caused by density variations within each aquifer (model layer) and differs between the stratified and continuous options to represent the density variation (fig. 2). The continuity of flow equation below surfaces 2 to N is

$$\delta_n \nabla\left(T_n^*\nabla \zeta_n\right) = n_e \frac{\partial \zeta_n}{\partial t} - \gamma_n + R_n \qquad n = 2,\dots,N,\qquad(11)$$

where

T_n^*	is the transmissivity between surface n and $N+1$ [L²/T],
δ_n	is the difference in the dimensionless density between surface n and $n+1$ [unitless], and
R_n	are pseudo-source terms below surface n [L/T].

For the stratified option (fig. 2B), δ_n for the upper zone ($n = 1$) is defined as

$$\delta_1 = v_1,\qquad(12)$$

where v_1 is the dimensionless density of zone 1. For zones n from 2 to N, δ_n is defined as

$$\delta_n = v_n - v_{n-1},\qquad(13)$$

where v_n is the dimensionless density of zone n. For the stratified option, R in equation 10 is defined as

$$R = -\sum_{n=1}^{N} \delta_n \nabla \left(T_n^* \nabla \zeta_n \right). \tag{14}$$

For the stratified option, R_n in equation 11 is defined as

$$R_n = -\nabla \left(T \nabla h \right) - \sum_{\substack{p=1 \\ p \neq n}}^{N} \delta_p \nabla \left(T_p^* \nabla \zeta_p \right) \qquad n = 2,3,\ldots N, \tag{15}$$

where

T_p^* is the transmissivity between surface p and $N+1$ [L²/T],

δ_p is the difference in the dimensionless density between surface p and $p+1$ [unitless], and

ζ_p is the elevation of surface p [L].

For the continuous density option (fig. 2C), the dimensionless density is defined for each surface n (μ_n), and δ_n is computed using equations 12 and 13, but with the dimensionless density of zone n calculated as the average of the defined dimensionless surface densities using

$$v_n = \frac{\mu_n + \mu_{n+1}}{2}. \tag{16}$$

For the continuous option, R in equation 10 is defined as

$$R = -\sum_{n=1}^{N} \delta_n \nabla \left(T_n^* \nabla \zeta_n \right) + \sum_{n=1}^{N} \varepsilon_n \nabla \left[T_n \nabla \left(\zeta_n - \zeta_{n+1} \right) \right], \tag{17}$$

where

ε_n is a measure of the variation of the dimensionless density over over zone n, and

T_n is the transmissivity of zone n [L²/T].

For the continuous option, R_n in equation 11 is defined as

$$R_n = \nabla \left(T \nabla h \right) \sum_{\substack{p=1 \\ p \neq n}}^{N} \delta_p \nabla \left(T_p^* \nabla \zeta_p \right) + $$

$$\sum_{p=n}^{N} \varepsilon_p \nabla \left[T_p \nabla \left(\zeta_p \quad \zeta_{p+1} \right) \right] \qquad n = 2,3,\ldots N, \tag{18}$$

where

ε_p is a measure of the variation of the dimensionless density over zone p, and

T_p is the transmissivity of zone p [L²/T].

ε_p is a result of the derivation of equation 18 (eqs. 28–30 in Bakker, 2003) and is defined as

$$\varepsilon_p = \frac{v_{p+1} - v_p}{6}, \tag{19}$$

where ε_p is a measure of the variation of the dimensionless density over zone p. When $\varepsilon_p = 0$ for all zones, equations 17 and 18 reduce to equations 14 and 15, used for the stratified option.

Alternative Derivation of Horizontal Vertically Integrated Variable-Density Groundwater Flow

To facilitate understanding of the mathematical formulation of the SWI2 Package, an alternative derivation of the SWI2 equations is developed and used to calculate flow in specific zones. These alternative equations are then manipulated within a dimensional context using finite differences to show their equivalence to the mathematical formulation previously described.

For an aquifer with a horizontal top and bottom, and having two distinct fluid types (fig. 4A) the discharge between two observation wells for zone 1, Q_1, is

$$Q_1 = -K \overline{b}_1 dy \frac{\left(h_b - h_a \right)}{dx}, \tag{20}$$

where

$\overline{b}_1$ is the average thickness of zone 1 between wells a and b [L],

dy is the aquifer width perpendicular to flow,

h_b is the freshwater head at the top of the aquifer at well b [L],

h_a is the freshwater head at the top of the aquifer at well a [L], and

dx is $x_b - x_a$ [L].

The discharge between two wells for zone 2 is

$$Q_2 = -K \overline{b}_2 dy \frac{\left(\hat{h}_{b,2} - \hat{h}_{a,2} \right)}{dx}, \tag{21}$$

where

$\overline{b}_2$ is the average thickness of zone 2 between wells a and b [L],

$\hat{h}_{b,2}$ is the calculated freshwater head for zone 2 at well b [L], and

$\hat{h}_{a,2}$ is the calculated freshwater head for zone 2 at well a [L].

The average freshwater head for zone 2 is one approach that could be used to calculate the specific discharge for zone 2. The average freshwater head for zone 2 at well b, $\overline{h}_{b,2}$, calculated using equation 5, is

$$\overline{h}_{b,2} = h_b + v_1 \left(\zeta_{b,1} - \zeta_{b,2} \right) + \frac{1}{2} v_2 \left(\zeta_{b,2} - \zeta_{b,3} \right). \tag{22}$$

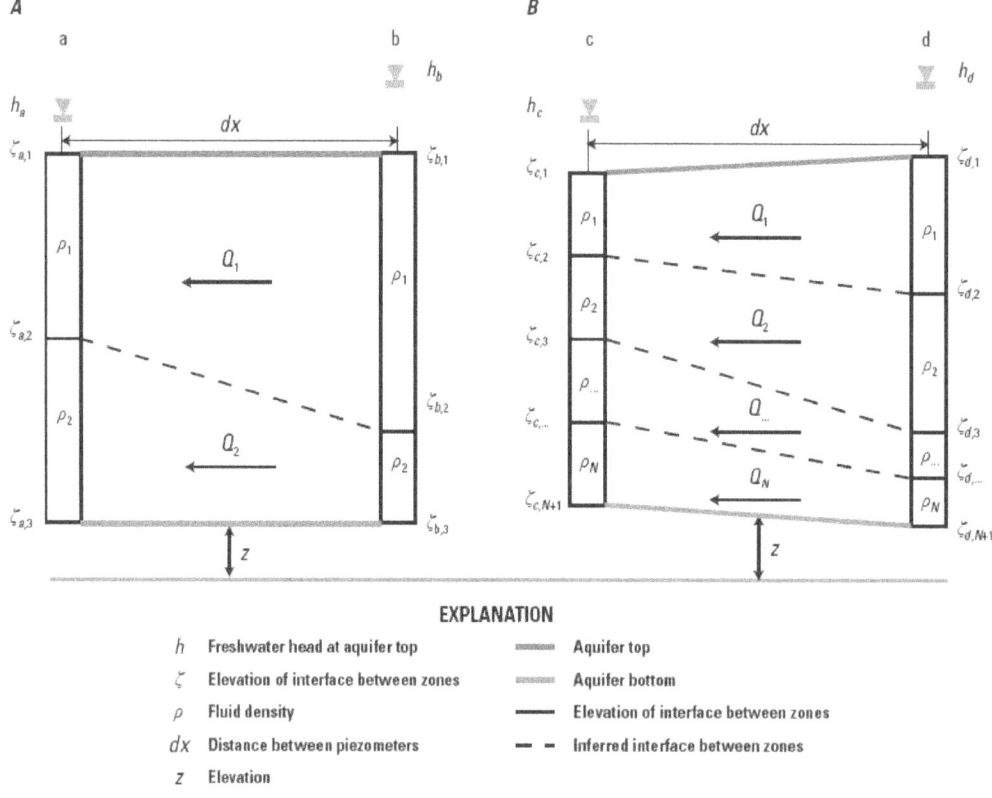

EXPLANATION

h	Freshwater head at aquifer top	▨▨▨▨	Aquifer top
ζ	Elevation of interface between zones	▨▨▨▨	Aquifer bottom
ρ	Fluid density	———	Elevation of interface between zones
dx	Distance between piezometers	– – –	Inferred interface between zones
z	Elevation		
Q	Zone groundwater flow		

Figure 4. Examples of *A,* a horizontal aquifer with two zones between wells a and b, and *B,* a sloping aquifer with two or more zones between wells c and d.

Similarly, the average freshwater head for zone 2 at well a, $\overline{h}_{a,2}$, is

$$\overline{h}_{a,2} = h_a + v_1\left(\zeta_{a,1} - \zeta_{a,2}\right) + \frac{1}{2}v_2\left(\zeta_{a,2} - \zeta_{a,3}\right). \quad (23)$$

The disadvantage of using the average freshwater head for zone 2 is that the freshwater heads are referenced to the vertical midpoint of zone 2; therefore, they are referenced to different elevations in wells a and b. This could lead to a nonzero discharge for no-flow conditions.

To overcome the deficiency of simply using average heads for zone 2, the heads are adjusted to the same vertical elevation. Modification of equation 5 to calculate the head at well a or b in zone 2 referenced to the same datum, under the assumption that hydrostatic conditions exist within an aquifer, results in

$$\hat{h}_{w,2} = \overline{h}_{w,2} + v_2\left(z_{w,2} - z_0\right), \quad (24)$$

where $\overline{h}_{w,2}$ is the average freshwater head for zone 2 at well a or b [L]. $z_{w,2}$ is the elevation of the vertical midpoint of zone 2 at well a or b [L], and z_0 is an arbitrary datum [L]. The value

of z_0 can have any value and, if set to zero, can be used to rewrite equation 22 as

$$\hat{h}_{b,2} = h_b + v_1\left(\zeta_{b,1} - \zeta_{b,2}\right) +$$
$$\frac{1}{2}v_2\left(\zeta_{b,2} - \zeta_{b,3}\right) + \frac{1}{2}v_2\left(\zeta_{b,2} + \zeta_{b,3}\right). \quad (25)$$

Equation 25 can be further simplified to

$$\hat{h}_{b,2} = h_b + v_1\left(\zeta_{b,1} - \zeta_{b,2}\right) + v_2\zeta_{b,2}. \quad (26)$$

Similarly, equation 23 can be modified and simplified to be

$$\hat{h}_{a,2} = h_a + v_1\left(\zeta_{a,1} - \zeta_{a,2}\right) + v_2\zeta_{a,2}. \quad (27)$$

By combining equations 20, 21, 26, and 27, the total discharge for the aquifer shown in figure 4*A* can be written as

$$Q = Q_1 + Q_2 = K\overline{b}dy\frac{\left(h_b - h_a\right)}{dx}$$
$$K\overline{b}_2 dy\frac{\left[v_1\left(\zeta_{b,1} - \zeta_{a,1}\right) - v_1\left(\zeta_{b,2} - \zeta_{a,2}\right) + v_2\left(\zeta_{b,2} - \zeta_{a,2}\right)\right]}{dx}, \quad (28)$$

where $\bar{b}$ is average aquifer thickness between wells a and b. The latter term in equation 28 represents a pseudo-source term, R, and may be written as

$$R = K\bar{b}_2 dy \frac{\left[\nu_1\left(\zeta_{b,1} - \zeta_{a,1} \right) - \left(\nu_1 - \nu_2 \right)\left(\zeta_{b,2} - \zeta_{a,2} \right) \right]}{dx}, \qquad (29)$$

which can be transformed into equation 14 using the difference in dimensionless density between surfaces, δ_n, defined in equations 12 and 13.

A two-zone example is given of pseudo-source term calculations for the horizontal aquifer (fig. 4A). In this example,

- wells a and b are 10 meters (m) apart,

- the aquifer has a hydraulic conductivity of 1 meter per day (m/d),

- the aquifer width perpendicular to flow is specified to be 1 m,

- the aquifer top is specified to be 10 m,

- the aquifer bottom is specified to be 0 m,

- $\nu_1 = 0.0$,

- $\nu_2 = 0.025$,

- the interface is specified to be 5 m above the base of the aquifer in well a,

- the interface is specified to be 2.5 m above the base of the aquifer in well b,

- $h_a = 11$ m, and

- $h_b = 20$ m.

The calculated discharge for these two zones together using aquifer properties and the freshwater heads h_a and h_b is -9.0 m³/d (cubic meters per day). The pseudo-source term calculated using equation 29 is 0.023 m³/d and adding this to the 9.0 m³/d discharge calculated using aquifer properties and freshwater heads results in a corrected discharge of -8.977 m³/d for the aquifer. The discharge for zone 1 and zone 2, calculated using equations 21, 26, and 27 are -5.625 and -3.352 m³/d, respectively. The sum of the discharge is equal to the corrected discharge calculated using equation 28. Finally, the corrected discharge calculated using equation 28 is equal to the discharge calculated using equation 10 and values of zero for storage, vertical discharge ($q_{z,t}$ and $q_{z,b}$), and source (γ) terms.

For aquifers having two or more density zones, equation 28 can be extended to incorporate additional surfaces. The discharge for zone n can be expressed as

$$Q_n = -K\bar{b}_n dy \frac{\left(\hat{h}_{b,n} - \hat{h}_{a,n} \right)}{dx}, \qquad (30)$$

where $\bar{b}_n$ is average thickness of zone n between wells a and b. As before, the average head in zone n at well b is calculated as

$$\bar{h}_{b,n} = h_b + \sum_{\substack{p\,1 \\ n>1}}^{n} \nu_p \left(\zeta_{b,p} - \zeta_{b,p+1} \right) + \frac{1}{2} \nu_n \left(\zeta_{b,n} - \zeta_{b,n+1} \right), \qquad (31)$$

where $\zeta_{b,p}$ is the elevation of surface p in well b and $\zeta_{b,n}$ is surface elevation of surface n in well b. Correcting equation 31 to the datum $z_0 = 0.0$ results in

$$\hat{h}_{b,n} = h_b + \sum_{\substack{p\,1 \\ n>1}}^{n} \nu_p \left(\zeta_{b,p} - \zeta_{b,p+1} \right) +$$

$$\frac{1}{2} \nu_n \left(\zeta_{b,n} - \zeta_{b,n+1} \right) + \frac{1}{2} \nu_n \left(\zeta_{b,n} + \zeta_{b,n+1} \right), \qquad (32)$$

which can be simplified to

$$\hat{h}_{b,n} = h_b + \sum_{\substack{p\,1 \\ n>1}}^{n} \nu_p \left(\zeta_{b,p} - \zeta_{b,p+1} \right) + \nu_n \zeta_{b,n}. \qquad (33)$$

Similarly, the calculated freshwater head for zone n at well a is

$$\hat{h}_{a,n} = h_a + \sum_{\substack{p\,1 \\ n>1}}^{n} \nu_p \left(\zeta_{a,p} - \zeta_{a,p+1} \right) + \nu_n \zeta_{a,n}, \qquad (34)$$

where $\zeta_{a,p}$ is the elevation of surface p in well a and $\zeta_{a,n}$ is the elevation of surface n in well a.

The total discharge for aquifers having two or more zones, such as those shown in figure 4B, can be calculated by summing equations 30, 33, and 34 for each zone. After rearrangement and simplification, the total discharge for aquifers having two or more zones is

$$Q = \sum_{n\,1}^{N} Q_n = -K\bar{b}dy \frac{\left(h_b - h_a \right)}{dx} -$$

$$\frac{Kdy}{dx} \sum_{n\,1}^{N} \bar{b}_n \left[\begin{array}{l} \sum_{\substack{p\,1 \\ n>1}}^{n} \nu_p \left(\zeta_{b,p} - \zeta_{a,p} \right) - \\ \sum_{\substack{p\,1 \\ n>1}}^{n} \nu_p \left(\zeta_{b,p+1} - \zeta_{a,p+1} \right) + \\ \nu_n \left(\zeta_{b,n} - \zeta_{a,n} \right) \end{array} \right]. \qquad (35)$$

Vertical Leakage Between Aquifers

The vertical leakage from aquifer $k+1$ to aquifer k is computed as the product of the head difference between the two aquifers and the vertical leakance. The head difference is the difference between the head at the top of aquifer $k+1$ and the head at the bottom of aquifer k, so that the vertical leakage, $q_{z,b}$, may be written as

$$q_{z,b} = L\left(h_{k+1} - h_{b,k}\right), \tag{36}$$

where

L is the leakance between aquifers k and $k+1$ [T^{-1}],

$h_{b,k}$ is the freshwater head at the bottom of aquifer k [L], and

h_{k+1} is the head at the top of aquifer $k+1$ [L].

The head at the bottom of aquifer k can be calculated using

$$h_{b,k} = h_k + \int_{\zeta_1}^{\zeta_{N+1}} \nu \, d\zeta = h_k + \sum_{n=1}^{N} \nu_n \left(\zeta_n - \zeta_{n+1}\right), \tag{37}$$

where h_k is the freshwater head at the top of aquifer k [L]. The vertical leakage from aquifer k to $k-1$, $q_{z,t}$, is also calculated using equations 36 and 37.

When water leaks vertically between aquifers, it is transferred to the same zone it is received from. Hence, freshwater is added to the freshwater zone and saltwater is added to the saltwater zone. Local areas may exist within a model where this is not possible. Four situations may be distinguished and are explained by considering the case of two aquifers with two zones (freshwater and seawater separated by an interface):

1. Seawater leaks upward into an aquifer that contains only freshwater. In this case, seawater is added to the seawater zone at the bottom of the overlying aquifer and the interface moves upward from the base of the overlying aquifer (as in example simulation 4).

2. Freshwater leaks upward into an aquifer that contains only seawater. In this case, the freshwater is converted to seawater and is added to the seawater zone in the overlying aquifer. This is also referred to as the instantaneous mixing model.

3. Freshwater leaks downward into an aquifer that contains only seawater. In this case, freshwater is added to the freshwater zone at the top of the underlying aquifer and the interface moves downward from the top of the underlying aquifer (as in example simulation 4).

4. Seawater leaks downward into a cell that contains only freshwater. In this case, the seawater is converted to freshwater and is added to the freshwater zone in the underlying aquifer. This is also referred to as the instantaneous mixing model.

Seawater Intrusion (SWI2) Package Implementation

SWI2 is designed such that a MODFLOW model for single-density flow may be converted to a seawater intrusion model by adding one input file to the model, and specifying the SWI2 input and output files in the MODFLOW NAME file. No changes are needed in the input files of any of the other packages. The only additional change that may be required is in how the coastal boundary is represented, as explained later. The formulation of the SWI2 Package and guidelines for the construction of a SWI2 model are described here.

Solution Procedure

Initial values must be specified for all dependent variables in each aquifer (model layer) and include (1) the freshwater head, h, at the saturated top of each aquifer (in the MODFLOW BAS file) and (2) the elevations ζ_n of surfaces 2 through N in each aquifer (in the MODFLOW SWI2 file). Boundary conditions may be specified using standard MODFLOW packages. During the simulation, the elevations of surfaces 2 through N will be calculated as part of the solution. Calculation of the surfaces depends on the flow into and out of each zone, which are calculated as part of the solution of the groundwater flow equation. If a density zone has a thickness of zero (the overlying and underlying surfaces have the same elevation), then flow into and out of that zone is zero. At the end of each SWI2 time step (which has a length that is less than or equal to the MODFLOW time step length), the program determines if a density zone with a non-zero thickness in one cell should begin to flow into the same density zone with zero thickness in an adjacent cell. This process for activating zero-thickness density zones is called the tip and toe tracking procedure and is described later.

Given values for head and elevations of surfaces at time t, the head and surface elevations at time $t+\Delta t$ are computed as follows. Equation 10 is solved to compute the head at time $t+\Delta t$, using the elevations of the surfaces at time t to compute the pseudo-source term, R. In case of multiaquifer flow, the heads in all aquifers are solved simultaneously, as is done in MODFLOW simulations without the SWI2 Package. Next, equation 11 is solved for each ζ_n surface elevation at time $t+\Delta t$. Because the vertical fluxes between model layers are known from the head solution, ζ_n surface elevations are solved separately for each aquifer (model layer).

Two options are available for computing the R and R_n terms in equations 10 and 11. In the first option, the default non-adaptive SWI2 time step option, R and R_n are computed using the head values at time $t+\Delta t$ and the ζ_n surface elevations at time t, which is equivalent to keeping the flow field fixed during a time step. In the second option, the optional adaptive SWI2 time step option ("ADAPTIVE" OPTION in dataset 1), R and R_n are computed using the head (HNEW) and ζ_n surface elevations (ZETA) values from the previous outer iteration, which is equivalent to the approach used by MODFLOW to solve for groundwater heads during a time step.

Finite-Difference Solution of the Groundwater Flow Equation

The groundwater flow equation 10 is solved using the standard MODFLOW finite difference approach. The model domain is discretized into NROW rows, NCOL columns, and NLAY layers. Both aquifers and confining units are treated in the same manner. The head in row i, column j, layer k is written as $h_{i,j,k}$. The discharge out of the block through the right face of the cell is called $Q_{i,j+1/2,k}$. The discharge out of the block through the front face of the cell is called $Q_{i+1/2,j,k}$. The discharge out of the block through the lower face of the cell is called $Q_{i,j,k+1/2}$. In a single-density model, the discharge through the lower face is computed with a central-in-space difference scheme as

$$Q_{i,j,k+1/2} = CV_{i,j,k+1/2}\left(h_{i,j,k} - h_{i,j,k+1}\right), \qquad (38)$$

where $CV_{i,j,k+1/2}$ is the vertical conductance in row i and column j between layers k and $k+1$. In SWI2, the head in layer k represents the freshwater head at the top of layer k, and the head in layer $k+1$ is the freshwater head at the top of layer $k+1$. With this formulation, the flux between layer k and $k+1$ is

$$Q_{i,j,k+1/2} = CV_{i,j,k+1/2}\left(h_{i,j,k} - h_{i,j,k+1} - BOUY_{i,j,k}\right), \quad (39)$$

where the term $BOUY_{i,j,k}$ is defined for the stratified and continuous options as

$$BOUY_{i,j,k} = \sum_{n=1}^{NZONES} NUS_n\left(ZETA_{i,j,k-1,n} - ZETA_{i,j,k-1,n+1}\right) + $$
$$\frac{1}{2}\left(ZETA_{i,j,k-1,NZONES+1} - ZETA_{i,j,k,1}\right) \times \qquad , (40)$$
$$\left(NUBOT_{i,j,k-1} - NUTOP_{i,j,k}\right)$$

where
NZONES	is the number of simulated fluid density zones within an aquifer (NSRF+1),
NUS	is the dimensionless density (v),
ZETA	is current ζ_n surface elevation,
NUBOT	is the dimensionless density at the bottom of the overlying layer, and
NUTOP	is the dimensionless density at the top of current layer.

The number of simulated zones (NZONES) is equal to one more than the total number of simulated surfaces within the aquifer (NSRF).

The MODFLOW finite difference equivalent of governing differential equation 10 is based on (5–1) of Harbaugh (2005) and modified to include the additional SWI2 pseudo-source term, R, and the flux correction term, $BOUY$, and yields

$$CR_{i,j-1/2,k}\left(h_{i,j-1,k}^m - h_{i,j,k}^m\right) + CR_{i,j+1/2,k}\left(h_{i,j+1,k}^m - h_{i,j,k}^m\right) +$$
$$CC_{i-1/2,j,k}\left(h_{i-1,j,k}^m - h_{i,j,k}^m\right) + CC_{i+1/2,j,k}\left(h_{i+1,j,k}^m - h_{i,j,k}^m\right) +$$
$$CV_{i,j,k-1/2}\left(h_{i,j,k-1}^m - h_{i,j,k}^m\right) + CV_{i,j,k+1/2}\left(h_{i,j,k+1}^m - h_{i,j,k}^m\right) +$$
$$P_{i,j,k}h_{i,j,k}^m + Q_{i,j,k} =$$
$$SS_{i,j,k}^m\left(DELR_j DELC_i DELV_{i,j,k}\right)\frac{h_{i,j,k}^m - h_{i,j,k}^{m-1}}{t^m - t^{m-1}} + \qquad , (41)$$
$$R_{i,j,k}^{m-1} - CV_{i,j,k-1/2}BOUY_{i,j,k-1}^{m-1} + CV_{i,j,k+1/2}BOUY_{i,j,k}^{m-1}$$

where
CR	is the conductance in the row direction,
h	is the simulated groundwater head in a cell,
m	is the MODFLOW time step number,
CC	is the conductance in the column direction,
CV	is the conductance in the layer (vertical) direction,
P	is the total conductance of all general head-dependent external source terms in a cell (see eq. 2–21 in Harbaugh, 2005),
Q	is the total specified general external source term (see eq. 2–21 in Harbaugh, 2005),
SS	is the specific storage of the cell,
DELR	is the width of column j,
DELC	is the width of row i, and
DELV	is the saturated thickness of the cell.

The superscripts $m-1$ and m indicate the previous and current time step number, respectively. If the optional adaptive SWI2 time step option ("ADAPTIVE" OPTION in dataset 1) is used, R and $BOUY$ from the previous outer iteration (see fig. 2–6 in Harbaugh, 2005) are used instead of the values from the previous $m-1$ time step.

The finite difference approximation of the pseudo-source term R for the stratified option in equation 14 is

$$R_{i,j,k}^{m-1} =$$
$$-\sum_{n=1}^{NZONES} DELNUS_n \begin{bmatrix} SWICUMCR_{i,j-1/2,k,n}\left(\begin{matrix}ZETA_{i,j-1,k,n}^{m-1} - \\ ZETA_{i,j,k,n}^{m-1}\end{matrix}\right) + \\ SWICUMCR_{i,j+1/2,k,n}\left(\begin{matrix}ZETA_{i,j+1,k,n}^{m-1} - \\ ZETA_{i,j,k,n}^{m-1}\end{matrix}\right) + \\ SWICUMCC_{i-1/2,j,k,n}\left(\begin{matrix}ZETA_{i-1,j,k,n}^{m-1} - \\ ZETA_{i,j,k,n}^{m-1}\end{matrix}\right) + \\ SWICUMCC_{i+1/2,j,k,n}\left(\begin{matrix}ZETA_{i+1,j,k,n}^{m-1} - \\ ZETA_{i,j,k,n}^{m-1}\end{matrix}\right) \end{bmatrix} , (42)$$

where

$DELNUS$ is the measure of the density variation in zone N (defined in equations 12 and 13 as δ_n),

$SWICUMCR$ is the cumulative conductance in the row direction below surface n, and

$SWICUMCC$ is the cumulative conductance in the row direction below surface n.

$SWICUMCR$ is computed as

$$SWICUMCR_{i,j+1/2,k,n} = \sum_{p}^{NZONES}{}_{n} SWICR_{i,j+1/2,k,p}, \quad (43)$$

where $SWICR$ is the conductance in the row direction for a zone. $SWICR$ is computed as

$$SWICR_{i,j+1/2,k,p} = CR_{i,j+1/2,k} \frac{THICKRF_{i,j+1/2,k,p}}{\sum_{n}^{NZONES}{}_{1} THICKRF_{i,j+1/2,k,n}}, \quad (44)$$

where $THICKRF$ is the linearly interpolated zone thickness at the interface between columns j and $j+1$ (right face). $THICKRF$ is calculated as

$$THICKRF_{i,j+1/2,k,p} =$$
$$\frac{DELR_{j+1}\left(ZETA_{i,j,k,p} - ZETA_{i,j,k,p+1}\right)}{\left(DELR_{j+1} + DELR_j\right)} +$$
$$\frac{DELR_j\left(ZETA_{i,j+1,k,p} - ZETA_{i,j+1,k,p+1}\right)}{\left(DELR_{j+1} + DELR_j\right)} . \quad (45)$$

The same process can be applied to the calculation of $SWICUMCC$ and results in

$$SWICUMCC_{i+1/2,j,k,n} = \sum_{p}^{NZONES}{}_{n} SWICC_{i+1/2,j,k,p}, \quad (46)$$

where $SWICC$ is the conductance in the column direction for a zone. $SWICC$ is calculated as

$$SWICC_{i+1/2,j,k,p} = CC_{i+1/2,j,k} \frac{THICKFF_{i+1/2,j,k,p}}{\sum_{n}^{NZONES}{}_{1} THICKFF_{i+1/2,j,k,n}}, \quad (47)$$

where $THICKFF$ is the linearly interpolated zone thickness at the interface between rows i and $i+1$ (front face). $THICKFF$ is calculated as

$$THICKFF_{i+1/2,j,k,p} =$$
$$\frac{DELC_{i+1}\left(ZETA_{i,j,k,p} - ZETA_{i,j,k,p+1}\right)}{\left(DELC_{i+1} + DELC_i\right)} +$$
$$\frac{DELC_i\left(ZETA_{i+1,j,k,p} - ZETA_{i+1,j,k,p+1}\right)}{\left(DELC_{i+1} + DELC_i\right)} . \quad (48)$$

The finite difference approximation of the pseudo-source term, R, for the continuous option in equation 17 is

$$R_{i,j,k}^{m-1} =$$

$$\sum_{n}^{NZONES}{}_{1} EPS_n \left[\begin{array}{l} SWICR_{i,j-1/2,k,n}\left[\left(\begin{array}{l}ZETA_{i,j-1,k,n}^{m-1} - \\ ZETA_{i,j-1,k,n+1}^{m-1}\end{array}\right) - \left(\begin{array}{l}ZETA_{i,j,k,n}^{m-1} - \\ ZETA_{i,j,k,n+1}^{m-1}\end{array}\right)\right] + \\ SWICR_{i,j+1/2,k,n}\left[\left(\begin{array}{l}ZETA_{i,j+1,k,n}^{m-1} - \\ ZETA_{i,j+1,k,n+1}^{m-1}\end{array}\right) - \left(\begin{array}{l}ZETA_{i,j,k,n}^{m-1} - \\ ZETA_{i,j,k,n+1}^{m-1}\end{array}\right)\right] + \\ SWICC_{i-1/2,j,k,n}\left[\left(\begin{array}{l}ZETA_{i-1,j,k,n}^{m-1} - \\ ZETA_{i-1,j,k,n+1}^{m-1}\end{array}\right) - \left(\begin{array}{l}ZETA_{i,j,k,n}^{m-1} - \\ ZETA_{i,j,k,n+1}^{m-1}\end{array}\right)\right] + \\ SWICC_{i+1/2,j,k,n}\left[\left(\begin{array}{l}ZETA_{i+1,j,k,n}^{m-1} - \\ ZETA_{i+1,j,k,n+1}^{m-1}\end{array}\right) - \left(\begin{array}{l}ZETA_{i,j,k,n}^{m-1} - \\ ZETA_{i,j,k,n+1}^{m-1}\end{array}\right)\right] \end{array} \right] -$$

$$\sum_{n}^{NZONES}{}_{1} DELNUS_n \left[\begin{array}{l} SWICUMCR_{i,j-1/2,k,n}\left(ZETA_{i,j-1,k,n}^{m-1} - ZETA_{i,j,k,n}^{m-1}\right) + \\ SWICUMCR_{i,j+1/2,k,n}\left(ZETA_{i,j+1,k,n}^{m-1} - ZETA_{i,j,k,n}^{m-1}\right) + \\ SWICUMCC_{i-1/2,j,k,n}\left(ZETA_{i-1,j,k,n}^{m-1} - ZETA_{i,j,k,n}^{m-1}\right) + \\ SWICUMCC_{i+1/2,j,k,n}\left(ZETA_{i+1,j,k,n}^{m-1} - ZETA_{i,j,k,n}^{m-1}\right) \end{array} \right] , \quad (49)$$

where *EPS* is the measure of the density variation in zone N (defined in eq. 19). Equation 49 is identical to the pseudo-source term for the stratified option (eq. 42) with the exception of the *EPS*, *SWICR*, and *SWICC* terms. *EPS* is zero for the stratified option and allows equation 49 to be programmatically used in SWI2 to solve for both the stratified and continuous options.

Finite-Difference Solution of Zeta Surfaces

Once a solution for the head in the aquifer is obtained for MODFLOW time step number m, the fluxes between layers are computed using equation 39. The fluxes between aquifers (model layers) are known after convergence of the groundwater flow equation (eq. 41) and the *ZETA* surface elevation for surface n at MODFLOW time step number m can be calculated for each layer separately. Movement of the *ZETA* surfaces is governed by differential equation 11. The form of equation 11 is similar to the differential equation solved by standard MODFLOW-2005 (eq. 10). The finite difference approximation of equation 11 for zone n is

$$
\begin{aligned}
&SWISOLCR_{i,j-1/2,k,n}(ZETA^m_{i,j-1,k,n} - ZETA^m_{i,j,k,n}) + \\
&SWISOLCR_{i,j+1/2,k,n}(ZETA^m_{i,j+1,k,n} - ZETA^m_{i,j,k,n}) + \\
&SWISOLCC_{i-1/2,j,k,n}(ZETA^m_{i-1,j,k,n} - ZETA^m_{i,j,k,n}) + \\
&SWISOLCC_{i+1/2,j,k,n}(ZETA^m_{i+1,j,k,n} - ZETA^m_{i,j,k,n}) - \\
&\frac{SSZ_{i,j,k}(DELR_j DELC_i)}{t^m - t^{m-1}} ZETA^m_{i,j,k,n} = \\
&-\frac{SSZ_{i,j,k}(DELR_j DELC_i)}{t^m - t^{m-1}} ZETA^{m-1}_{i,j,k,n} - G_{i,j,k,n} + R^{m-1}_{i,j,k,n}
\end{aligned}
\tag{50}
$$

where
SWISOLCR   is the conductance in the row direction used to solve for *ZETA* surface n,
SWISOLCC   is the conductance in the column direction used to solve for *ZETA* surface n,
SSZ   is the effective porosity,
G   is the source term below surface n, and
R   is the pseudo-source term below surface n.
SWISOLCR   is calculated as

$$
SWISOLCR_{i,j+1/2,k,n} = DELNUS_n SWICUMCR_{i,j+1/2,k,n} - EPS_p SWICR_{i,j+1/2,k,n}.
\tag{51}
$$

Recall that *EPS* is zero for the stratified option (ISTRAT=1). Similarly, *SWISOLCC* is calculated as

$$
\begin{aligned}
SWISOLCC_{i+1/2,j,k,n} = \\
DELNUS_n SWICUMCC_{i+1/2,j,k,n} - \\
EPS_p SWICC_{i+1/2,j,k,n}.
\end{aligned}
\tag{52}
$$

The known source term, $G_{i,j,k,n}$, is calculated as

$$
\begin{aligned}
G_{i,j,k,n} = &\left(-Q_{i,j,k,n} - SS_{i,j,k} DELR_j DELC_i \frac{h^{m-1,}_{i,j,k}}{t^m - t^{m-1}}\right) - \\
&\left(P_{i,j,k,n} - \frac{SS_{i,j,k} DELR_j DELC_i}{t^m - t^{m-1}}\right) h^m_{i,j,k} + \\
&\left(QZ_{i,j,k-1/2,n} - QZ_{i,j,k+1/2,n}\right)
\end{aligned}
\tag{53}
$$

and is composed of terms that are known after the groundwater flow equation (eq. 41) is solved. The terms within the first parenthesis of equation 53 are the known *RHS* flux terms (*RHSPRESWI*) prior to adding the pseudo-source term, *R*. The terms in the second parenthesis of equation 53 are the boundary and storage conductances used to calculate the groundwater head, h, portion of the boundary and storage flux term, *HCOF*. The *QZ* terms in the third parenthesis of equation 53, are the vertical leakage source terms from the same zone in overlying and underlying model layers. The sum of all of the terms in equation 53 represent the net boundary, storage, and vertical flux for the current MODFLOW time step number, m, and is applied to the boundary zone (ISOURCE) defined for each model cell. The boundary zone options available in SWI2 are discussed in more detail in the following section.

For the stratified option, R_n in equation 50 is calculated as

$$
R^{m-1}_{i,j,k,n} = -\begin{bmatrix} SWICUMCR_{i,j-1/2,k,n}\left(h^m_{i,j-1,k} - h^m_{i,j,k}\right) + \\ SWICUMCR_{i,j+1/2,k,n}\left(h^m_{i,j+1,k} - h^m_{i,j,k}\right) + \\ SWICUMCC_{i-1/2,j,k,n}\left(h^m_{i-1,j,k} - h^m_{i,j,k}\right) + \\ SWICUMCC_{i+1/2,j,k,n}\left(h^m_{i+1,j,k} - h^m_{i,j,k}\right) \end{bmatrix} - \\
\sum^{NZONES}_{\substack{p\,1 \\ p \neq n}} DELNUS_p \begin{bmatrix} SWICUMCR_{i,j-1/2,k,p}\left(\begin{matrix} ZETA^{m-1}_{i,j-1,k,p} - \\ ZETA^{m-1}_{i,j,k,p} \end{matrix}\right) + \\ SWICUMCR_{i,j+1/2,k,p}\left(\begin{matrix} ZETA^{m-1}_{i,j+1,k,p} - \\ ZETA^{m-1}_{i,j,k,p} \end{matrix}\right) + \\ SWICUMCC_{i-1/2,j,k,p}\left(\begin{matrix} ZETA^{m-1}_{i-1,j,k,p} - \\ ZETA^{m-1}_{i,j,k,p} \end{matrix}\right) + \\ SWICUMCC_{i+1/2,j,k,p}\left(\begin{matrix} ZETA^{m-1}_{i+1,j,k,p} - \\ ZETA^{m-1}_{i,j,k,p} \end{matrix}\right) \end{bmatrix} \cdot
\tag{54}
$$

For the continuous option, R_n in equation 50 is calculated as

$$
R_{i,j,k,n}^{m-1} = -\begin{bmatrix} SWICUMCR_{i,j-1/2,k,n}\left(h_{i,j-1,k}^{m} - h_{i,j,k}^{m}\right) + \\ SWICUMCR_{i,j+1/2,k,n}\left(h_{i,j+1,k}^{m} - h_{i,j,k}^{m}\right) + \\ SWICUMCC_{i-1/2,j,k,n}\left(h_{i-1,j,k}^{m} - h_{i,j,k}^{m}\right) + \\ SWICUMCC_{i+1/2,j,k,n}\left(h_{i+1,j,k}^{m} - h_{i,j,k}^{m}\right) \end{bmatrix} -
$$

$$
\sum_{\substack{p\,1\\p\neq n}}^{NZONES} DELNUS_p \begin{bmatrix} SWICUMCR_{i,j-1/2,k,p}\begin{pmatrix} ZETA_{i,j-1,k,p}^{m-1} - \\ ZETA_{i,j,k,p}^{m-1} \end{pmatrix} + \\ SWICUMCR_{i,j+1/2,k,p}\begin{pmatrix} ZETA_{i,j+1,k,p}^{m-1} - \\ ZETA_{i,j,k,p}^{m-1} \end{pmatrix} + \\ SWICUMCC_{i-1/2,j,k,p}\begin{pmatrix} ZETA_{i-1,j,k,p}^{m-1} - \\ ZETA_{i,j,k,p}^{m-1} \end{pmatrix} + \\ SWICUMCC_{i+1/2,j,k,p}\begin{pmatrix} ZETA_{i+1,j,k,p}^{m-1} - \\ ZETA_{i,j,k,p}^{m-1} \end{pmatrix} \end{bmatrix} +
$$

$$
\sum_{p\,n}^{NZONES} EPS_p \begin{bmatrix} SWICR_{i,j-1/2,k,p}\begin{bmatrix} \begin{pmatrix} ZETA_{i,j-1,k,p}^{m-1} - \\ ZETA_{i,j,k,p}^{m-1} \end{pmatrix} - \\ \begin{pmatrix} ZETA_{i,j-1,k,p+1}^{m-1} - \\ ZETA_{i,j,k,p+1}^{m-1} \end{pmatrix} \end{bmatrix} + \\ SWICR_{i,j+1/2,k,p}\begin{bmatrix} \begin{pmatrix} ZETA_{i,j+1,k,p}^{m-1} - \\ ZETA_{i,j,k,p}^{m-1} \end{pmatrix} - \\ \begin{pmatrix} ZETA_{i,j-1,k,p+1}^{m-1} - \\ ZETA_{i,j,k,p+1}^{m-1} \end{pmatrix} \end{bmatrix} + \\ SWICC_{i-1/2,j,k,p}\begin{bmatrix} \begin{pmatrix} ZETA_{i-1,j,k,p}^{m-1} - \\ ZETA_{i,j,k,p}^{m-1} \end{pmatrix} - \\ \begin{pmatrix} ZETA_{i-1,j,k,p+1}^{m-1} - \\ ZETA_{i,j,k,p+1}^{m-1} \end{pmatrix} \end{bmatrix} + \\ SWICC_{i+1/2,j,k,p}\begin{bmatrix} \begin{pmatrix} ZETA_{i+1,j,k,p}^{m-1} - \\ ZETA_{i,j,k,p}^{m-1} \end{pmatrix} - \\ \begin{pmatrix} ZETA_{i+1,j,k,p+1}^{m-1} - \\ ZETA_{i,j,k,p+1}^{m-1} \end{pmatrix} \end{bmatrix} \end{bmatrix} . \qquad (55)
$$

Equation 55 is identical to the pseudo-source term for the stratified option (eq. 54), with the exception of the EPS, $SWICR$, and $SWICC$ terms. EPS is zero for the stratified option and allows equation 55 to be programmatically used in SWI2 to solve for the ζ surfaces with both the stratified and continuous options.

Sinks and Sources

Many of the existing MODFLOW packages that add or subtract water from model cells may be applied to SWI2 models. There are two considerations that must be taken into account for source and sink boundary packages. The first consideration relates to the type of head that is specified for all head-dependent packages. In SWI2, all boundary heads must be specified as freshwater heads at the top of the aquifer (model layer), which may or may not be referenced to sea level. For example, the heads specified for the GHB and RIV Packages must be the freshwater head at the top of the layer.

The second consideration relates to the type of water (freshwater or seawater, for example) that flows into the aquifer from the boundary packages. The type of water for sinks and sources is specified in the SWI2 input file through the ISOURCE parameter. The water type for sinks and sources must correspond to one of the density zones defined for the model. The ISOURCE value is used to define the correct zone to apply to known boundary condition source terms, G_n, in equation 50. Each cell has one value for ISOURCE. Three options are available for specifying the water types of sinks and sources:

1. ISOURCE > 0. Sources and sinks are of the same type as water in zone ISOURCE. If such a zone is not present in the cell, sources and sinks interact with the zone at the top of the aquifer.

2. ISOURCE = 0. Sources and sinks are of the same type of water as the water at the top of the aquifer.

3. ISOURCE < 0. Sources are of the same type as water in zone ISOURCE. Sinks are the same water type as the water at the top of the aquifer. This option is useful for modeling the ocean bottom where sources of water are saltwater and groundwater sinks (submarine groundwater discharge) discharge water from the zone at the top of the aquifer. For example, if a cell with a general head boundary (GHB) was assigned an ISOURCE value of -2 in a model with one active interface (NSRF=1) and two zones, water that infiltrates into the aquifer from the GHB cell would be seawater (zone 2), whereas water that flows out of the cell would be of the same type as the water at the top of the aquifer (and may not be seawater).

Tip and Toe Tracking

At the end of every time step, it must be determined whether the boundaries of the surfaces that separate the zones move horizontally. The boundaries of each surface are flux-specified during a time step; as such, the surfaces can move up or down along the boundary during a time step. The boundary of a surface is either near the bottom of an aquifer, referred to as the toe, or near the top of an aquifer, referred to as a tip. The algorithm used to handle the toe, discussed next, operates separately along the rows and columns for each zone in each layer of the MODFLOW model.

Consider surface $ZETA_{i,j,k,n}$ shown in figure 5. At the end of a time step, boundary cell j contains the toe. Cell j has a horizontal length of $DELR_j$, a bottom elevation of $BOTM_{i,j,k}$, and an effective porosity of $SSZ_{i,j,k}$ (fig. 5); the elevation of the surface in the toe cell is called $ZETA_{i,j,k,n}$. The adjacent empty cell, $j+1$, has a horizontal length of $DELR_{j+1}$, a bottom elevation of $BOTM_{i,j+1,k}$, and effective porosity of $SSZ_{i,j+1,k}$.

The general principle of the toe tracking algorithm is that the toe is moved into the adjacent empty cell when the slope of the surface at the toe exceeds a user-specified maximum toe slope, i_{max} (TOESLOPE). The toe is moved from cell j to $j+1$ when

$$i_{i,j+1/2,k,n} = \frac{ZETA_{i,j,k,n} - BOTM_{i,j+1,k}}{\frac{1}{2}\left(DELR_j + DELR_{j+1}\right)} > i_{max}. \qquad (56)$$

When the simulated toe slope, $i_{i,j+1/2,k,n}$, is too steep, the surface for zone n, in toe cell j is lowered a small amount, $\Delta ZETA_{i,j,k,n}$, and the surface for zone n is introduced in adjacent empty cell $j+1$, such that the new slope is $(1-\alpha)i_{max}$, where α (ALPHA) is a coefficient having a default value of 0.1. The toe is moved from cell j to $j+1$ if

$$ZETA_{i,j,k,n} - BOTM_{i,j+1,k} > \Delta ZETA_{max} =$$
$$\frac{1}{2}(DELR_{i,j,k} + DELR_{i,j+1,k})i_{max}, \qquad (57)$$

then $ZETA_{i,j,k,n}$ should be lowered by $\Delta ZETA_{i,j,k,n}$, calculated as

$$\Delta ZETA_{i,j,k,n} =$$
$$\alpha \frac{SSZ_{i,j+1,k}DELR_{j+1}}{SSZ_{i,j,k}DELR_j + SSZ_{i,j+1,k}DELR_{j+1}} \Delta ZETA_{max}, \qquad (58)$$

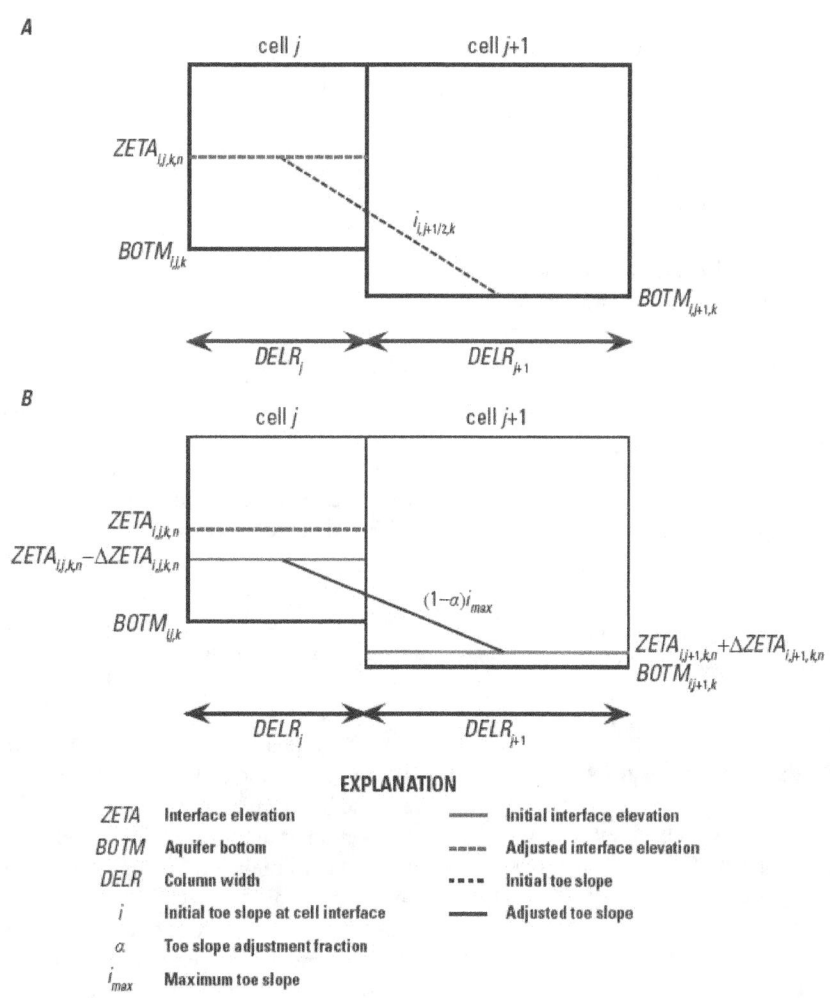

EXPLANATION

ZETA	Interface elevation	——	Initial interface elevation
BOTM	Aquifer bottom	----	Adjusted interface elevation
DELR	Column width	····	Initial toe slope
i	Initial toe slope at cell interface	——	Adjusted toe slope
α	Toe slope adjustment fraction		
i_{max}	Maximum toe slope		

Figure 5. Variables used in the toe tracking algorithm. Similar parameters are used in the tip tracking algorithm.

Similarly, $ZETA_{i,j+1,k,n}$ should be raised above the base, $BOTM_{i,j+1,k}$, by $\Delta ZETA_{i,j+1,k,n}$, which is calculated as

$$\Delta ZETA_{i,j+1,k,n} = \\ \alpha \frac{SSZ_{i,j,k} DELR_j}{SSZ_{i,j,k} DELR_j + SSZ_{i,j+1,k} DELR_{j+1}} \Delta ZETA_{\max}. \quad (59)$$

The toe tracking algorithm can produce undesired results when the aquifer bottom slope is larger than the user-defined toe slope (`TOESLOPE`).

When the surface for zone n in a toe cell ($j+1$, fig. 5) is below a minimum depth threshold level, the toe is moved into the adjacent non-empty cell. The minimum depth threshold is calculated as

$$\left(ZETA_{i,j+1,k,n} \right)_{\min} = \beta \Delta ZETA_{i,j+1,k,n}, \quad (60)$$

where β (`BETA`) is a coefficient having a default value of 0.1.

Application of the algorithm for a tip is identical to that for the toe, with the exception that the top of the model layer, $BOTM_{i,j,k-1}$, is used in equations 56 and 57. $ZETA_{i,j,k,n}$ would be increased by $\Delta ZETA_{i,j+1,k,n}$, and $ZETA_{i,j+1,k,n}$ would be decreased by $\Delta ZETA_{i,j+1,k,n}$.

In general, tip and toe slopes should be based on representative interface slopes (known or estimated) at the tip and toe of the aquifer, respectively. In cases where the interface slopes at the tip and toe of the aquifer are not known, the dimensionless density difference between freshwater and seawater and the head gradient at the coast and the toe of the interface can be used to estimate reasonable values.

Simulation of Aquifers and Confining Units

Aquifers and confining units need to be explicitly modeled as regular layers. As a result, quasi-three-dimensional layers, specified using non-zero `LAYCB` values for one or more model layers (Harbaugh, 2005, p. 2–16), should not be used in models using the SWI2 Package. Example simulation 3 demonstrates the suggested approach for explicitly representing thin confining units.

Interface Movement Between Layers

During pumping within a model layer, it is possible for a $ZETA$ surface to be pulled upward through the bottom of the layer from the model layer beneath it. Conversely, when the volume of a freshwater lens is growing, for example because of artificial recharge, it is possible for a $ZETA$ surface to be pushed through the bottom of a model layer into the underlying layer. This vertical movement of $ZETA$ surfaces into overlying and (or) underlying model layers is correctly simulated by the SWI2 Package, provided cell sizes and time

steps are small enough. Equation 53 accounts for vertical movement of surfaces, implicitly during numerical solution of $ZETA$ surfaces, in cells where fluid of one density is moving into overlying or underlying cells actively simulating the same zone (that is, cells with $ZETA$ surface n for zone p is not at the top or bottom of the model layer).

For cases in which fluid of a specific density is at the bottom of cell i,j,k-1 and at the top of the underlying cell i,j,k, and an upward flux exists (from layer k to k-1), the increase in the $ZETA$ surface in cell i,j,k-1 is calculated as

$$ZETA^m_{i,j,k-1,n} = ZETA^{m-1}_{i,j,k-1,n} + \\ \frac{DELT^m \cdot \left[CV_{i,j,k-1/2} \left(h^m_{i,j,k} - h^m_{i,j,k-1} - BOUY^{m-1}_{i,j,k-1} \right) \right]}{DELR_j \cdot DELC_i \cdot SSZ_{i,j,k-1}}, \quad (61)$$

where $DELT$ is the time-step length between time step m-1 and m. For cases with a downward flux, the decrease in the $ZETA$ surface in cell i,j,k is calculated as

$$ZETA^m_{i,j,k,n} = ZETA^{m-1}_{i,j,k,1} + \\ \frac{DELT^m \cdot \left[CV_{i,j,k-1/2} \left(h^m_{i,j,k} - h^m_{i,j,k-1} - BOUY^{m-1}_{i,j,k-1} \right) \right]}{DELR_j \cdot DELC_i \cdot SSZ_{i,j,k}}, \quad (62)$$

where $ZETA_{i,j,k,1}$ is the top of cell i,j,k.

When water leaks upward from a cell, it is possible that less dense water (for example, freshwater) leaks upward into an overlying cell with less dense water of the same type at the top of the cell and denser water (for example, seawater) at the bottom of the cell. In this case, the less dense water is added as a source term to the equivalent density zone in the overlying cell and the interface for the less dense water will move downward. This may be interpreted to mean that fingers grow infinitely fast. Similarly, it is possible that denser water (for example, seawater) leaks downward into an underlying cell with less dense water (for example, freshwater) at the top of the cell and more dense water of the same type at the bottom of the cell. Using the same approach, the denser water from the overlying cell is added as a source term to the equivalent density zone in the underlying cell and the interface for the denser water will move upward.

In cases where denser water leaks upward into a cell only containing less dense water, the denser water is added as a source term to the equivalent density zone in the overlying cell and the interface for the denser water will move upward from the bottom of the cell; this is also referred to as upconing (as in example simulation 4). Similarly, where less dense water leaks downward into a cell only containing denser water, the less dense water is added as a source term to the equivalent density zone in the overlying cell and the interface for the less dense water will move downward from the top of the cell.

In some cases, it is possible that less dense water leaks upward into the overlying cell, but only denser water is present in the cell above it. In such cases, the density of the less dense water increases to that of the denser water, which is also referred to as the instantaneous mixing model. Similarly, it is possible for denser water to leak downward into a cell containing only less dense fluid cell. In this case, the density of the denser water decreases to that of the less dense water. Example simulation 4 demonstrates the vertical movement of an interface in response to freshwater recharge and groundwater withdrawals with the SWI2 Package.

Simulating Coastal Boundaries

Coastal boundaries generally need to be simulated using multiple model grid cells extending offshore into the ocean. In general, it is recommended that general head boundaries (GHB) be used to represent offshore freshwater heads at the ocean bottom. In cases where the ocean bottom does not extend into the ocean (for example, where a rocky cliff is present offshore), additional high hydraulic conductivity cells should be used to represent the ocean; this approach is similar to the one used for modeling lakes using high hydraulic conductivity cells. The model must be extended below or into the ocean because the tip cell of the last surface cannot be at the boundary of the model. If the model does not extend below or into the ocean, saltwater cannot enter the model and the volume of saltwater in the model domain will remain constant; this is an incorrect representation of coastal aquifers. Example simulations 3, 4, 6, and 7 demonstrate the suggested approach for representing coastal boundaries.

Adaptive SWI2 Time Steps

An optional adaptive time stepping algorithm has been implemented to dynamically adjust the number of SWI2 time steps in a MODFLOW time step. The algorithm has no effect on the lengths of MODFLOW time steps and stress periods. Use of this option will increase the number of SWI2 time steps used during a MODFLOW time step when the tip or toe movement exceeds a user-defined value. The number of SWI2 time steps for each MODFLOW time step is increased if in the row-direction:

$$ADPTVAL = \frac{ADPTFCT\left(top_{i,j,k}^{m} - ZETA_{i,j,k,n}^{m}\right)}{TIPSLOPE\dfrac{DELR_j + DELR_{j\pm1}}{2}} > 1, \quad (63)$$

where

$ADPTVAL$	is the SWI2 time step adjustment factor,
$ADPTFCT$	is the user-defined tip toe factor,
top	is the top of model layer k ($BOTM_{i,j,k-1}$) or the simulated head ($HNEW_{i,j,k}$) if the head is below the top of the model layer, and
$TIPSLOPE$	is the user-defined maximum tip slope (i_{max}).

A user-defined $ADPTFCT$ value of 2.0 has been observed to work well for the test problems evaluated. Equation 63 is

evaluated in the column-direction by substituting $DELC_i$ and $DELC_{i\pm1}$ for $DELR_j$ and $DELR_{j\pm1}$.

Similarly, the number of SWI2 time steps for a MODFLOW time step is increased if in the row-direction:

$$ADPTVAL = \frac{ADPTFCT\left(ZETA_{i,j,k,n}^{m} - BOTM_{i,j,k}\right)}{TOESLOPE\dfrac{DELR_j + DELR_{j\pm1}}{2}} > 1, \quad (64)$$

where $TOESLOPE$ is the user-defined maximum toe slope (i_{max}). $DELC_i$ and $DELC_{i\pm1}$ are substituted for $DELR_j$ and $DELR_{j\pm1}$ in equation 64 to evaluate the change in the toe in the column-direction.

The number of SWI2 time steps for a MODFLOW time step is increased for any cell with a non-zero zone thickness if

$$ADPTVAL = \frac{ADPTFCT\left(ZETA_{i,j,k,n}^{m} - ZETA_{i,j,k,n}^{m-1}\right)}{top_{i,j,k}^{m} - BOTM_{i,j,k}} > 1. \quad (65)$$

If $ADPTVAL$ in equation 63, 64, or 65 is greater than 1, the number of SWI2 time steps in a MODFLOW time step is increased using

$$NADPT = \min\left(NADPTMX, NADPT \cdot ADPTVAL\right), \quad (66)$$

where

$NADPT$	is the number of SWI2 time steps in the current MODFLOW time step, and
$NADPTMX$	is the user-defined maximum number of SWI2 time steps in any MODFLOW time step.

The number of SWI2 time steps in a MODFLOW time step are decreased if the number of SWI2 time steps in a MODFLOW time step was not increased in the previous MODFLOW time step. The number of SWI2 time steps per MODFLOW time step are decreased using

$$NADPT = \max\left(NADPTMN, \frac{NADPT}{2}\right), \quad (67)$$

where $NADPTMN$ is the user-defined minimum number of SWI2 time steps in any MODFLOW time step. $NADPT$ is only decreased in the first outer iteration of a MODFLOW time step. The adaptive SWI2 time step option can be used to simulate a fixed number of SWI2 time steps for each MODFLOW time step by setting $NADPTMN$ equal to $NADPTMX$. Setting $NADPTMN$ equal to $NADPTMX$ disables use of equations 66 and 67 in the SWI2 Package but allows for more refined movement of the $ZETA$ surfaces because the tip and toe can move multiple cells during a MODFLOW time step.

If the optional adaptive SWI2 time step option ("ADAP-TIVE" OPTION in dataset 1) is used, $ZETA$ terms from the previous outer iteration are used instead of terms from the previous MODFLOW time step (m-1) in the groundwater flow equation (eq. 41) and the *ZETA* surface equation (eq. 50). Integrated SWI2 pseudo-source terms are used in the groundwater flow equation (eq. 41). Incremental SWI2 terms, for each SWI2 time step, are used in the *ZETA* surface equation (eq. 50).

Using the SWI2 Package

The main input of the SWI2 file is the initial position of the *ZETA* surfaces. An elevation for each surface needs to be specified for every cell in the model. The *ZETA* surface is placed at the top of a model layer when a value is entered that is above the top of the model layer and, similarly, the *ZETA* surface is placed at the bottom of a model layer when a value is entered that is below the bottom of the model layer. For the case of a surface that is present at only one point in the vertical everywhere, the same grid of *ZETA* values may be entered for every model layer and SWI2 will determine in which cells the elevation of the *ZETA* surface falls between the top and model of each layer (see also example simulation 3). It is important to start with a reasonably smooth variation of the *ZETA* surface because it is not physically reasonable for a surface to be discontinuous or have sharp changes in gradient. When an irregular *ZETA* surface is entered, additional time in a transient simulation may be required before a physically realistic *ZETA* elevation is achieved in each cell, and it may take small time steps to reach an accurate solution.

Groundwater flow simulations using the SWI2 Package may be steady-state or transient, as specified in the discretization (DIS) file. Simulation of the change in the *ZETA* surface elevations and density (salinity) distribution is always transient. A steady-state head simulation combined with a transient change in the position of the *ZETA* surfaces is often reasonable, because the *ZETA* surfaces commonly react much slower to a change in the system than the head. When the head is simulated as steady-state, it means that aquifer storage is neglected and the heads respond instantaneously to a change in source and sink functions (for example, a well that begins pumping). An additional benefit of using the steady-state head option is that larger time steps may be used than when the transient option is used. It is important to recognize that the steady-state head option does not mean that the steady-state density distribution is computed. The steady-state density distribution can only be computed by running the simulation until the density distribution is no longer changing.

As mentioned in the section on sources and sinks, the head values entered in the input files for head-dependent source and sink packages must be specified as the freshwater head at the top of the aquifer (model layer), which may be referenced to an arbitrary datum other than sea level. Furthermore, the water type for sources and sinks must also be specified. The source and sink water type is specified using the ISOURCE input array in the SWI2 input file.

Output Files and Post Processing

Output files from a SWI2 simulation consist of standard MODFLOW output plus the SWI2 *ZETA* output file. The SWI2 *ZETA* output file includes the position of every active *ZETA* surface of every model layer for the same time steps for which heads are recorded. The format of the SWI2 *ZETA* output file is the same as that of a standard MODFLOW budget file. Heads and drawdowns are written in terms of freshwater heads at the top of each model layer. Therefore, to compare simulated model heads to observation data, observed head data should be converted to freshwater heads, or simulated heads should be converted to heads calculated using the ambient groundwater density at the well screen (point water heads). Optionally, *ZETA* surface observations at specified layer, row, and column locations can be saved to an ASCII or binary output file.

There are two main differences in the standard MODFLOW output files that are unique to SWI2. First, the summary listing (LIST) file records the volumetric budget for the entire model, as is recorded for regular MODFLOW models. In addition, the file records the volumetric budget for each zone. The volumetric budget for the entire model includes one new term, SWIADDTOCH, which needs to be added to the CONSTANT HEAD term to obtain the correct value and is only non-zero if the constant head cell is located in a cell where the ISOURCE term is not 1. The volumetric budget of each zone contains five terms. The first term, BOUNDS + STORAGE, is the volume change caused by flow to or from model boundaries plus volume changes due to changes in aquifer storage. "Model boundaries" is the cumulative term for all sinks and sources in the model, including but not limited to recharge, wells, and general head boundaries. The second term, CONSTANT HEAD, is the volume change caused by flow to or from constant head boundaries. The third term, ZONE CHANGE, is the volume change due to changes in the size of the zone. The inflow caused by ZONE CHANGE represents the decrease in the volume of the zone whereas the outflow caused by ZONE CHANGE represents the increase in the volume of the zone. The fourth term, ZONE CHG TIP/TOE, is the volume change caused by the tip/toe algorithm. The fifth term, ZONE MIXING, is the volume change due to instantaneous mixing of one zone with another as a result of vertical flow between layers in cases where denser water in one layer directly overlies fresher water in the underlying layer.

Second, the flow terms in the cell-by-cell flow file are expressed as volumetric fluxes rather than mass fluxes and represent the total flux in the model layer. The fluxes in the cell-by-cell flow file need to be adjusted by the terms SWIADDTOFLF, SWIADDTOFRF, and SWIADDTOFFF, for the lower face (LF), right face (RF), and front face (FF), respectively. These values are written to the cell-by-cell flow file or a separate file when the appropriate flag (ISWIBD > 0) is set in the SWI2 input file. Separate fluxes need to be computed for each zone to be able to do particle tracking. Such computations can be carried out using the information in the cell-by-cell flow file and the position of the *ZETA* surfaces, but they are currently not provided.

Tips for Designing MODFLOW-2005 Models Using the SWI2 Package

A good approach to start a SWI2 model is to start with a cross-sectional model for interface flow (one *ZETA* surface), keeping in mind that confining units must be modeled explicitly rather than implicitly (using a quasi-three-dimensional approach). If the intent is to model only one aquifer, then the SWI2 cross-sectional model may be one dimensional; for multiple aquifers, the cross-sectional model would be two-dimensional. The cross-sectional model for interface flow may be used to obtain insight into the flow pattern and evaluate appropriate grid resolution, aquifer parameters, hydrologic stresses, computer run times, and solver limitations. Once the cross-sectional model yields reasonable results, the model can be extended to two dimensions for a one aquifer model or to three dimensions for a multiaquifer model. Additional surfaces may also be added, again in a stepwise fashion.

A first estimate of an appropriate grid resolution may be obtained using existing guidelines for single-density models (for example, Anderson and Woessner, 1992). SWI2 computes the vertical movement of each *ZETA* surface only for *ZETA* surface elevations between the top and the bottom of the aquifer, so an additional constraint for SWI2 simulations is the horizontal grid resolution must be small enough to have at least a few *ZETA* surfaces with elevations between the top and bottom of the aquifer.

The choice of the time step is another important issue, and an iterative approach is recommended to determine an accurate time step. The user is advised to start with short simulations having short time steps and increase the simulation time and time step length when results indicate it is realistic to do so. A good initial choice for time step length is one that does not result in a change in *ZETA* surface elevation by more than 20 percent of the aquifer thickness during each iteration. An additional consideration in the selection of time step length is that the tip or toe can move laterally only one cell during each time step. If the time step is too large, the tip or toe may not move into adjacent cells as quickly as necessary, which results in a steep *ZETA* surface at the tip or toe. It is generally advisable to start a simulation with small time steps to gradually reduce any coarseness in the initial variation of the *ZETA* surface specified by the user. The sensitivity of the model to time steps lengths should be checked periodically during the model calibration process as adjustments to boundary conditions and aquifer properties can affect *ZETA* surface movement.

Tip and toe slopes need to be specified for SWI2 simulations and should be based on representative interface slopes (known or estimated) at the tip and toe of the aquifer, respectively. If the interface is expected to intrude 2,000 m in an aquifer that is 40 m thick, a reasonable value for the tip and toe slopes is the average slope of the interface, 0.02 in this case. In situations where it is difficult to estimate the interface slope, reasonable values may be obtained by dividing an estimate of the head gradient by the dimensionless density difference. For most real systems, head gradients are on the order of 1 meter per kilometer (0.001 meter per meter) and tip and toe slopes are therefore on the order of 40 meters per kilometer (0.04 meter per meter), using a dimensionless density difference of 0.025 based on the density of saltwater [1,025 kilograms per cubic meter (kg/m^3)]. When the resistance to outflow into the sea is small, the interface slope at the tip is probably larger than the slope at the toe, as in example simulation 3. When the resistance to outflow is larger, however, the slope at the tip may be smaller than the slope at the toe, as in example simulation 4. Extensive experimentation has shown that simulation results are not sensitive to tip and toe slope values. Therefore, it is advisable to use the same values for the tip and toe slopes initially and only differentiate between them when modeling results warrant such a change.

The adaptive time step algorithm implemented in SWI2 can also be useful for determining an appropriate time step for subsequent SWI2 simulations. Initially, the adaptive SWI2 time step algorithm can be used to determine the SWI2 time step necessary to meet the desired *ADAPTFCT* value for movement of the tips and toes. After determining the number of SWI2 time steps needed to satisfy *ADAPTFCT*, the number of MODFLOW time steps in each stress period can be increased to match the number of SWI2 time steps determined dynamically by the adaptive SWI2 time step algorithm. The number of SWI2 time steps per MODFLOW time step for each stress period are written to the MODFLOW LST file as a summary table near the end of the file to facilitate setting final MODFLOW time steps based on interim adaptive SWI2 time step simulations.

The SWI2 results that are typically evaluated at the end of a simulation include the ASCII output listing file and binary files that contain freshwater heads at the top of each model layer, position of each *ZETA* surface in each model layer, and cell-by-cell flows. Users should remember that flow lines will be perpendicular to freshwater-head contours only in a horizontal plane. Because the computed heads are the freshwater heads at the top of a model layer, flow will only be normal to contours of this head when the top of the aquifer is horizontal. Evaluation of *ZETA* surface observations can also be useful for evaluating and identifying instabilities in SWI2 results; instabilities can occur with large MODFLOW time steps, high aquifer hydraulic conductivities, or large boundary condition conductance values.

Most sinks and sources in a seawater intrusion model have the salinity of freshwater (for example, natural recharge) or seawater (for example, infiltration at the ocean bottom). Sources of brackish water typically do not exist in a seawater-intrusion model unless a brackish water source, such as an underlying aquifer not explicitly simulated, is present. In a real physical system, additional brackish water may be created through mixing processes such as dispersion and diffusion. In the absence of brackish sources, the amount of brackish water generally decreases during a simulation as the brackish water discharges to model boundaries because mixing processes are not represented in a SWI2 model.

Example Simulations

Six hypothetical and one regional model example are used to demonstrate the capabilities of the SWI2 Package. Example simulations 1-3 are two-dimensional cross-section simulations that are compared to existing solutions and (or) other numerical simulators, such as SEAWAT. Example simulation 4 is a three-dimensional island simulation with areal recharge that evaluates upconing of the freshwater-saltwater interface in response to groundwater withdrawals. Example simulation 5 is a multilayer radial upconing simulation that is compared to SEAWAT. Example simulation 6 evaluates the difference between SWI2 and SEAWAT when (1) density inversions occur, (2) dispersive or diffusive mixing is important, or (3) differences between horizontal and vertical hydraulic conductivity are large. The last example simulation, 7, is an application of the SWI2 Package in a large model used to simulate regional flow in the unconfined, water-table aquifer in Cape Cod, Massachusetts.

An abbreviated Name file (NAM), Discretization (DIS), Basic (BAS), Layer Property Flow (LPF), and SWI2 input datasets for example simulation 1 have been included in appendix 2, and can be used as a guide for developing MOD-FLOW datasets that include the SWI2 Package. All seven example simulations are available online at *http://water.usgs. gov/software/*.

Example 1: Rotating interface

Example 1 simulates transient movement of a freshwater-seawater interface separating two density zones in a two-dimensional vertical plane. The problem domain is 250 m long, 40 m high, and 1 m wide. The aquifer is confined, storage changes are not considered (all MODFLOW stress periods are steady-state), and the top and bottom of the aquifer are horizontal and impermeable (fig. 6A).

The domain is discretized into 50 columns, 1 row, and 1 layer, with respective cell dimensions of 5 m (DELR), 1 m (DELC), and 40 m. A constant head of 0 m is specified for column 50. The hydraulic conductivity is 2 m/d and the effective porosity (SSZ) is 0.2. A flow of 1 m^3/d of seawater is specified in column 1 and causes groundwater to flow from left to right in the model domain.

The domain contains one freshwater zone and one seawater zone, separated by an active *ZETA* surface, ζ_2, between the zones (NSRF=1) that approximates the 50-percent seawater salinity contour. A 400-day period is simulated using a constant time step of 2 days. Fluid density is represented using the stratified option (ISTRAT=1) and the elevation of

the interface is output every 100 days (every 50 time steps). The densities, ρ, of the freshwater and saltwater are 1,000 and 1,025 kg/m^3, respectively. Hence, the dimensionless densities, v, of the freshwater and saltwater are 0.0 and 0.025, respectively (computed with eq. 2). The maximum slope of the toe and tip is specified as 0.2 (TOESLOPE=TIPSLOPE=0.2), and default tip/toe parameters are used (ALPHA=BETA=0.1). Initially, the interface is at a 45° angle from $(x,z) = (80,0)$ to $(x,z) = (120,-40)$ (fig. 6A). The source/sink terms (ISOURCE) are specified to be freshwater everywhere (ISOURCE=1) except in cell 1 where saltwater enters the model and ISOURCE equals 2.

A comparison of results for SWI2 and the exact Dupuit solution of Wilson and Sa Da Costa (1982) are shown in figure 6B. The constant flow from left to right results in an average velocity of 0.125 m/d. The exact Dupuit solution is a rotating straight interface of which the center moves to the right with this velocity.

The volumetric budget for each zone is recorded in the MODFLOW listing (LIST) file. At the end of the simulation, the cumulative inflow into zone 2 (saltwater) consists of 400 m^3 of saltwater inflow from the boundary (inflow on the left side of the model) plus a ZONE CHANGE of 2.52 m^3 representing "inflow" caused by the interface moving downward and tip and toe algorithm zone changes of 1.65 m^3. The total outflow of zone 2 consists of a ZONE CHANGE of 402.5 m^3 representing "outflow" caused by the interface moving upward and tip and toe algorithm zone changes of 1.55 m^3. The total increase in the volume of the saltwater zone is 399.9 m^3, which is computed by subtracting the sum of the inflow ZONE CHANGE and tip and toe algorithm zone changes (ZONE CHG TIP/TOE) from the sum of outflow ZONE CHANGE and tip and toe algorithm zone changes (ZONE CHG TIP/TOE). The total increase in the volume of the saltwater zone is equal to the total inflow of saltwater from the boundary minus the error for zone 2 (0.097 m^3 or 0.02 percent of the total inflow).

Example 2: Rotating Brackish Zone

Example 2 is a modification of the rotating interface problem (example 1), and includes three zones and no inflow at the boundary. The problem domain is 300 m long, 40 m high and 1 m wide. The aquifer is confined, storage changes are not considered (because all MODFLOW stress periods are steady-state), and the top and bottom of the aquifer are horizontal and impermeable. A constant head of 0 m is specified at $x = 0$ m (column 1). Example 2 aquifer properties are identical to those used in Example 1.

The domain is discretized into 60 columns, 1 row and 1 layer, with respective cell dimensions of 5 m (DELR) 1 m (DELC) and 40 m. A period of 2,000 days is simulated using a constant time step of 2 days.

The groundwater is divided into three zones, freshwater, brackish, and seawater, that have dimensionless densities, v, of 0, 0.0125, and 0.025, respectively; the zones are separated by

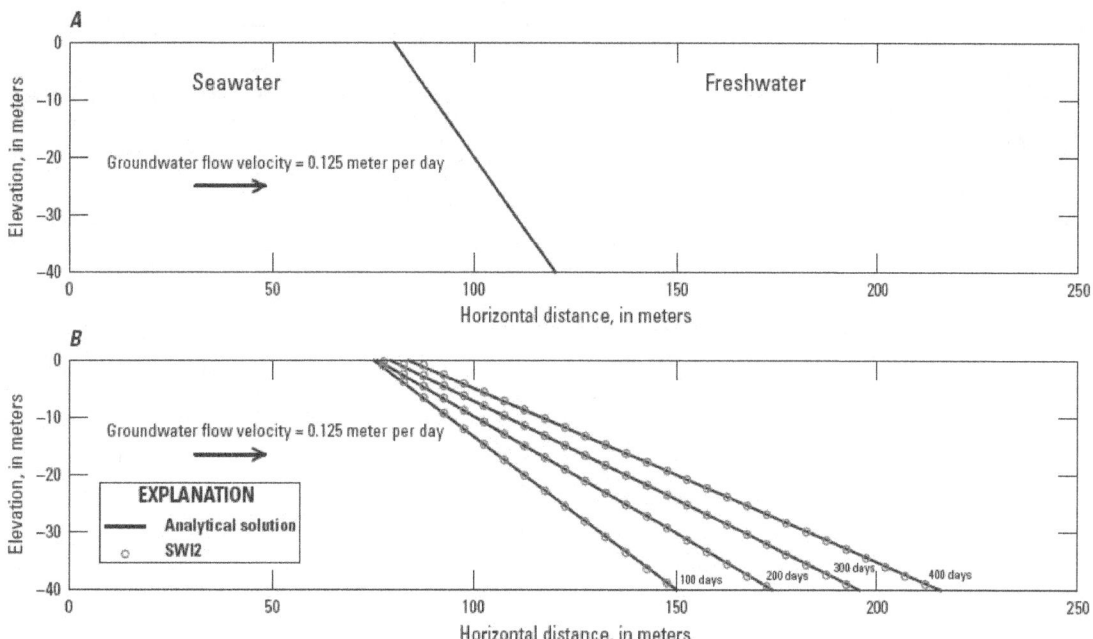

Figure 6. *A,* Initial freshwater-seawater interface elevation for example simulation 1 and ambient groundwater flow velocity, and *B,* a comparison of the freshwater-seawater interface elevation simulated with SWI2 and the exact solution at model cell centers at 100, 200, 300, and 400 days.

two active *ZETA* surfaces that represent the 25- and 75-percent seawater salinity contours (NSRF=2). Fluid density is represented using the stratified option (ISTRAT=1). The maximum slope of the toe and tip is specified as 0.4, and default tip and toe parameters are used (ALPHA=BETA=0.1). At time $t = 0$, both interfaces are straight and oriented 45° from horizontal. Initial *ZETA* surfaces 1 and 2 extend from $(x,z) = (150,0)$ to $(x,z) = (190,-40)$, and from $(x,z) = (110,0)$ to $(x,z) = (150,-40)$, respectively (fig. 7*A*). The brackish zone rotates toward a horizontal position over time.

A comparison of SWI2 and SEAWAT (Langevin and others, 2008) results at $t = 2,000$ days is shown in figure 7*B*. For SEAWAT, the aquifer is discretized into 80 layers, 1 row, and 300 columns (24,000 cells), with each cell having dimensions of 1×0.5 m in the horizontal and vertical directions, respectively. The number of cells in the SEAWAT model is 400 times as large as the number of cells in the SWI model. Velocity-dependent dispersion and diffusion are ignored in the transport component of SEAWAT (MT3DMS). As a result, the transport equation is solved using the total variation diminishing (TVD) scheme to reduce numerical dispersion. The brackish zone rotates slightly faster in SWI2 than in SEAWAT. Differences between SWI2 and SEAWAT results are a result of the use of the Dupuit approximation in SWI2 and numerical dispersion in SEAWAT, respectively. For example 2, run times for the SWI2 and SEAWAT simulations were approximately 0.4 and 240.0 seconds, respectively.

For illustrative purposes, the same problem is solved with the continuous option (ISTRAT=0), which requires

modification of the NU variable in item 4. Because three zones are simulated in Example 2, the NU variable now needs 4 input values (NSRF+2). The 4 NU values represent the dimensionless density at the top of zone 1, the dimensionless densities of the two surfaces, and the dimensionless density at the bottom of zone 3. The modified NU values for example 2 using the continuous option are

0.000000 0.000000 0.025000 0.025000,

and results in an average fluid density value equal to 50-percent seawater salinity ($v = 0.0125$) in the brackish zone. For this example problem, the difference between the positions of the surfaces after 2,000 days using the stratified and continuous options is small (fig. 7*C*).

Example 3: Freshwater-Seawater Interface Movement in a Two-Aquifer Coastal Aquifer System

Example 3 simulates transient movement of the freshwater-seawater interface in response to changing freshwater inflow in a two-aquifer coastal aquifer system. The problem domain is 4,000 m long, 41 m high, and 1 m wide. Both aquifers are 20 m thick and are separated by a leaky layer 1 m thick. The aquifers are confined, storage changes are not considered (all MODFLOW stress periods are steady-state), and the top and bottom of each aquifer is horizontal. The top of the upper aquifer and bottom of the lower aquifer are impermeable.

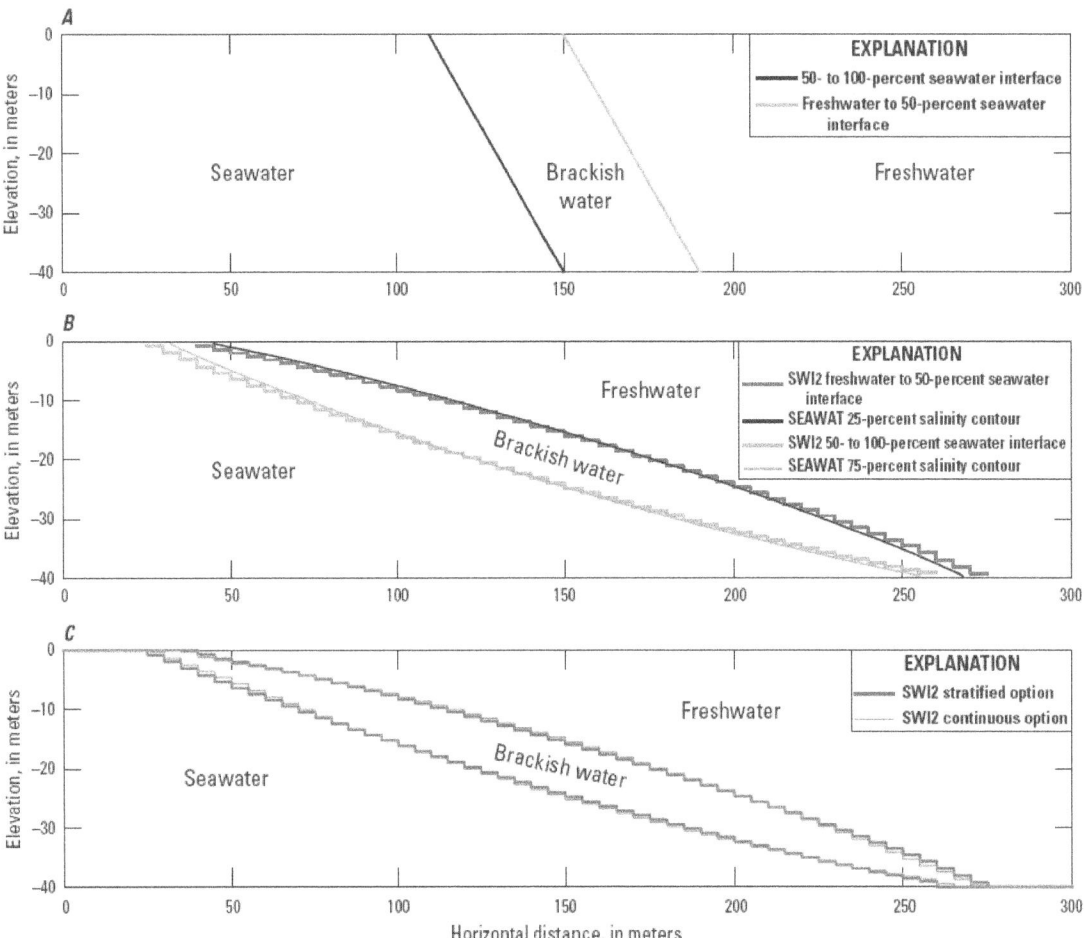

Figure 7. *A*, initial freshwater to 50-percent seawater and 50- to 100-percent seawater interface elevations for example simulation 2; *B*, the position of the brackish zone after 2,000 days, SWI2 calculated freshwater to 50-percent seawater and 50- to 100-percent seawater interface elevations are compared to SEAWAT 25- and 75-percent seawater salinity contours; and *C*, comparison of SWI2 results with the stratified and continuous option after 2,000 days.

The domain is discretized into 200 columns that are each 20 m long (DELR), 1 row that is 1 m wide (DELC), and 3 layers that are 20, 1, and 20 m thick. A total of 2,000 years are simulated using two 1,000-year stress periods and a constant time step of 2 years. The hydraulic conductivity of the top and bottom aquifer are 2 and 4 m/d, respectively, and the horizontal and vertical hydraulic conductivity of the confining unit are 1 and 0.01 m/d, respectively. The effective porosity is 0.2 for all model layers.

The left 600 m of the model domain extends offshore and the ocean boundary is represented as a general head boundary condition (GHB) at the top of model layer 1. A freshwater head of 0 m is specified at the ocean bottom in all general head boundaries. The GHB conductance that controls outflow from the aquifer into the ocean is 0.4 square meter per day (m²/d) and corresponds to a leakance of 0.02 d⁻¹ (or a resistance of 50 days).

The groundwater is divided into a freshwater zone and a seawater zone, separated by an active *ZETA* surface, ζ_2, between the zones (NSRF=1) that approximates the 50-percent seawater salinity contour. Fluid density is represented using the stratified density option (ISTRAT=1). The dimensionless densities, *ν*, of the freshwater and saltwater are 0.0 and 0.025. The tip and toe tracking parameters are a TOESLOPE of 0.02 and a TIPSLOPE of 0.04, a default ALPHA of 0.1, and a default BETA of 0.1. Initially, the interface between freshwater and seawater is straight, is at the top of aquifer 1 at $x = -100$, and has a slope of -0.025 m/m (fig. 8). The SWI2 ISOURCE parameter is set to -2 in cells having GHBs so that water that infiltrates into the aquifer from the GHB cells is saltwater (zone 2), whereas water that flows out of the model at the GHB cells is of the same type as the water at the top of the aquifer. In all other cells, the SWI2 ISOURCE parameter is set to 0, indicating boundary conditions have water that is identical to water at the top of the aquifer.

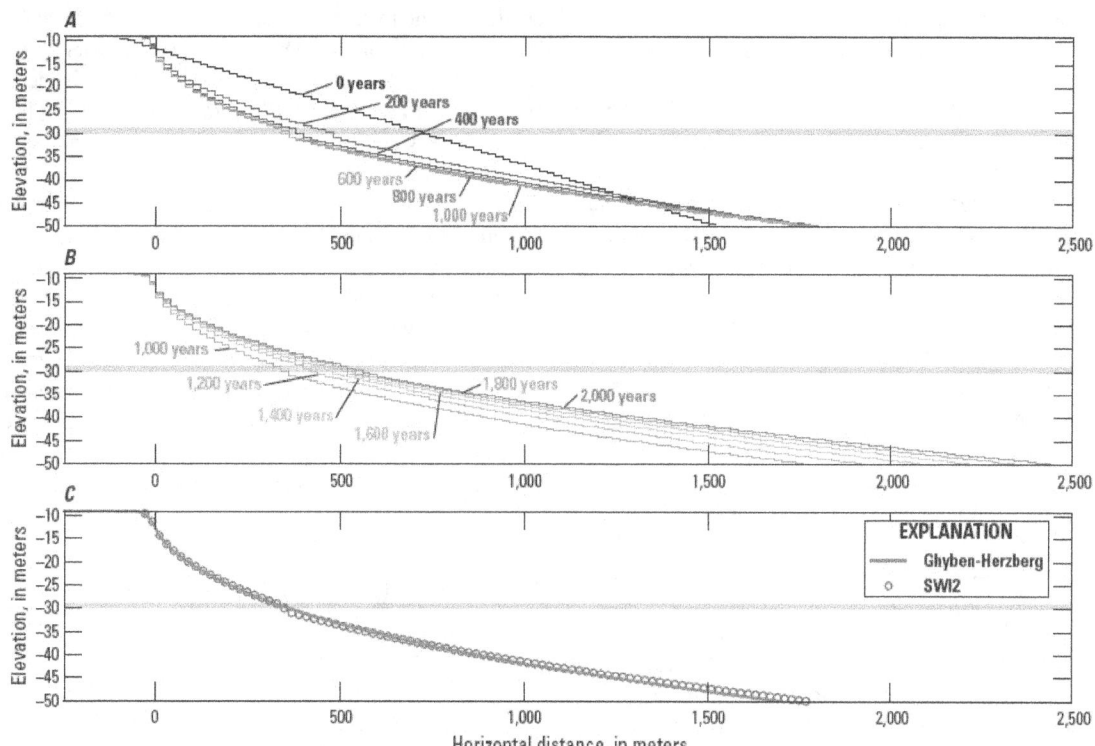

Figure 8. Simulated freshwater-seawater interface elevation for example simulation 3 (model domain extends from x = -600 to x = 3,400 meters): *A,* The position of the interface every 200 years after starting from the initial interface position (0 years). *B,* The position of the interface every 200 years after reducing the outflow to the coast by 50 percent from the value in *A,* and starting from the position at 1,000 years. *C,* Comparison of the SWI2 simulated freshwater-seawater interface after 1,000 years and the steady-state position based on the Ghyben-Herzberg relation and freshwater heads in the upper aquifer (model layer 1) after 1,000 years.

Initially, the net freshwater inflow rate of 0.03 m³/d specified at the right boundary causes flow to occur towards the ocean. The flow in each layer is distributed in proportion to the aquifer transmissivities. During the first 1,000-year stress period, a freshwater source of 0.01 m³/d is specified in the right-most cell (column 200) of the top aquifer, and a freshwater source of 0.02 m³/d is specified in the right-most cell (column 200) of the bottom aquifer. During the second 1,000-year stress period, these values are halved to reduce the net freshwater inflow to 0.015 m³/d, which is distributed in proportion to the transmissivities of both aquifers at the right boundary.

The initial and simulated interface positions during the first stress period are shown at 200-year increments in figure 8*A;* after 1,000 years, the interface approaches its steady-state position. Landward movement of the interface (to the right) caused by the reduction in freshwater flow towards the coast is shown at 200-year increments in figure 8*B.* In this case, the interface has not achieved a steady-state position in response to the reduction in freshwater flow toward the coast after 1,000 years at the end of the simulation. The confining unit affects the position of the interface in the upper and lower aquifers. The effect of vertical resistance to flow, which can be

related to different confining unit thicknesses and (or) different confining unit hydraulic properties, is evaluated in example simulation 6. The steady-state position, computed using the Ghyben-Herzberg relation (Ghyben, 1889; Herzberg, 1901) and the freshwater head in the upper aquifer, is comparable to SWI2 results after 1,000 years (fig. 8*C*).

Example 4: Upconing below a Pumping Well in a Two-Aquifer Island System

Example 4 simulates transient movement of the freshwater-seawater interface beneath an island in response to recharge and groundwater withdrawals. The island is 2,050×2,050 m and consists of two 20-m thick aquifers that extend below sea level. The aquifers are confined, storage changes are not considered (all MODFLOW stress periods are steady-state), and the top and bottom of each aquifer is horizontal. The top of the upper aquifer and the bottom of the lower aquifer are impermeable.

The domain is discretized into 61 columns, 61 rows, and 2 layers, with respective cell dimensions of 50 m (DELR), 50 m (DELC), and 20 m. A total of 230 years is simulated using three stress periods with lengths of 200, 12, and 18 years, with constant time steps of 0.2, 0.1, and 0.1 years, respectively.

The horizontal and vertical hydraulic conductivity of both aquifers are 10 m/d and 0.2 m/d, respectively. The effective porosity is 0.2 for both aquifers.

The model is extended 500 m offshore along all sides and the ocean boundary is represented as a general head boundary condition (GHB) in model layer 1. A freshwater head of 0 m is specified at the ocean bottom in all general head boundaries. The GHB conductance that controls outflow from the aquifer into the ocean is 62.5 m²/d and corresponds to a leakance of 0.025 d⁻¹ (or a resistance of 40 days).

The groundwater is divided into a freshwater zone and a seawater zone, separated by an active *ZETA* surface, ζ_2, between the zones (NSRF=1) that approximates the 50-percent seawater salinity contour. Fluid density is represented using the stratified density option (ISTRAT=1). The dimensionless density difference between freshwater and saltwater is 0.025. The tip and toe tracking parameters are a TOESLOPE and TIPSLOPE of 0.005, a default ALPHA of 0.1, and a default BETA of 0.1. Initially, the interface between freshwater and saltwater is 1 m below land surface on the island and at the top of the upper aquifer offshore. The SWI2 ISOURCE parameter is set to -2 in cells having GHBs so that water that infiltrates into the aquifer from the GHB cells is saltwater (zone 2), whereas water that flows out of the model at the GHB cells is identical to water at the top of the aquifer. ISOURCE in layer 2, row 31, column 36 is set to 2 so that a saltwater well may be simulated in the third stress period of simulation 2. In all other cells, the SWI2 ISOURCE parameter is set to 0, indicating boundary conditions have water that is identical to water at the top of the aquifer and can be either freshwater or saltwater, depending on the elevation of the active ZETA surface, ζ_2, in the cell.

A constant recharge rate of 0.4 millimeters per day (mm/d) is used in all three stress periods. The development of the freshwater lens is simulated for 200 years, after which a pumping well having a withdrawal rate of 250 m³/d is started in layer 1, row 31, column 36. For the first simulation (simulation 1), the well pumps for 30 years, after which the interface almost reaches the top of the upper aquifer layer. In the second simulation (simulation 2), an additional well withdrawing saltwater at a rate of 25 m³/d is simulated below the freshwater well in layer 2 , row 31, column 36, 12 years after the freshwater groundwater withdrawal begins in the well in layer 1. The saltwater well is intended to prevent the interface from upconing into the upper aquifer (model layer).

A cross section showing results along the centerline of the island through the wells is shown in figure 9. The position of the interface is shown at 40-year increments for stress period 1, during development of the freshwater lens (fig. 9A). After 200 years, the interface approaches the steady-state position. The position of the interface for simulation 1 is shown at 6-year increments in figure 9B. In simulation 1, the interface moves into the upper layer after approximately 14 years. The

position of the interface for simulation 2 is shown at 6-year increments in figure 9C. During the first 12 years of groundwater withdrawals, the position of the interface is identical for simulations 1 and 2. After 12 years, the saltwater well begins pumping saltwater from the lower aquifer (model layer 2) and the progression of the upconing of the interface is reduced. The position of the interface in the aquifer over time is shown for simulations 1 and 2 in the cells containing the freshwater and saltwater wells (row 31, column 36) in figure 9D. Simulation results are identical during the first 12 years, because groundwater withdrawals are the same in both simulations. Addition of the saltwater well causes an initial decrease in the interface elevation and an eventual increase in the interface elevation, although the rate of upward movement of the interface is less in the simulation with both the freshwater and saltwater wells than the simulation with only the freshwater well.

Example 5: Radial Upconing Problem

Example 5 simulates transient movement of the freshwater-seawater interface in response to groundwater withdrawals and is based on the problem of Zhou and others (2005). The aquifer is 120-m thick, confined, storage changes are not considered (because all MODFLOW stress periods are steady-state), and the top and bottom of the aquifer are horizontal and impermeable.

The domain is discretized into a radial flow domain, centered on a pumping well, having 113 columns, 1 row, and 6 layers, with respective cell dimensions of 25 m (DELR), 1 m (DELC), and 20 m. A total of 8 years is simulated using two 4-year stress periods and a constant 1-day time step.

The horizontal and vertical hydraulic conductivity of the aquifer are 21.7 and 8.5 m/d, respectively, and the effective porosity of the aquifer is 0.2. The hydraulic conductivity and effective porosity values applied in the model are based on the approach of logarithmic averaging of interblock transmissivity (LAYAVG=1) to better approximate analytical solutions for radial flow problems (Langevin, 2008). Constant heads were specified at the edge of the model domain (column 113) in all 6 layers, with specified freshwater heads of 0.0 m (fig. 10).

The groundwater is divided into a freshwater zone and a seawater zone, separated by an active *ZETA* surface, ζ_2, between the zones (NSRF=1) that approximates the 50-percent seawater salinity contour. Fluid density is represented using the stratified density option (ISTRAT=1). The dimensionless density difference between freshwater and saltwater is 0.025. The tip and toe tracking parameters are a TOESLOPE and TIPSLOPE of 0.025, a default ALPHA of 0.1, and a default BETA of 0.1. Initially, the interface between freshwater and saltwater is at an elevation of -100 m (fig. 10). The SWI2 ISOURCE parameter is set to 1 in the constant head cells in layers 1-5 and to 2 in the constant head cell in layer 6. This ensures that inflow from the constant head cells in model layers 1-5 and 6 is freshwater and saltwater, respectively. In

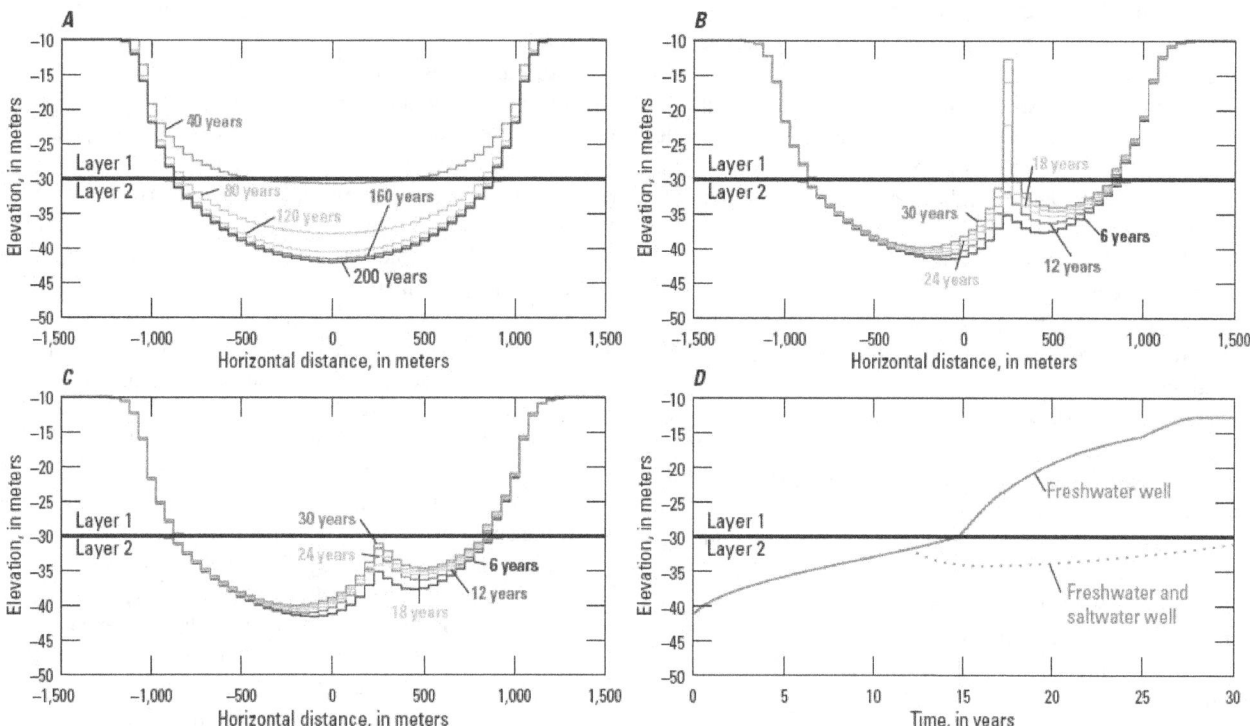

Figure 9.   Example simulation 4 results along centerline of island along row 31 through wells located in column 36: *A*, The position of the freshwater-seawater interface every 40 years during recharge conditions. *B*, The position of the freshwater-seawater interface every 6 years in response to groundwater withdrawals from a freshwater well in model layer 1. *C*, The position of the freshwater-seawater interface every 6 years in response to groundwater withdrawals from a freshwater well in model layer 1 and a saltwater well withdrawing saltwater from model layer 2 after 12 years. *D*, Elevation of interface at horizontal location of the freshwater and saltwater wells (row 31 column 36) as a function of time for the simulations without the saltwater well in model layer 2 and with saltwater well started after 12 years.

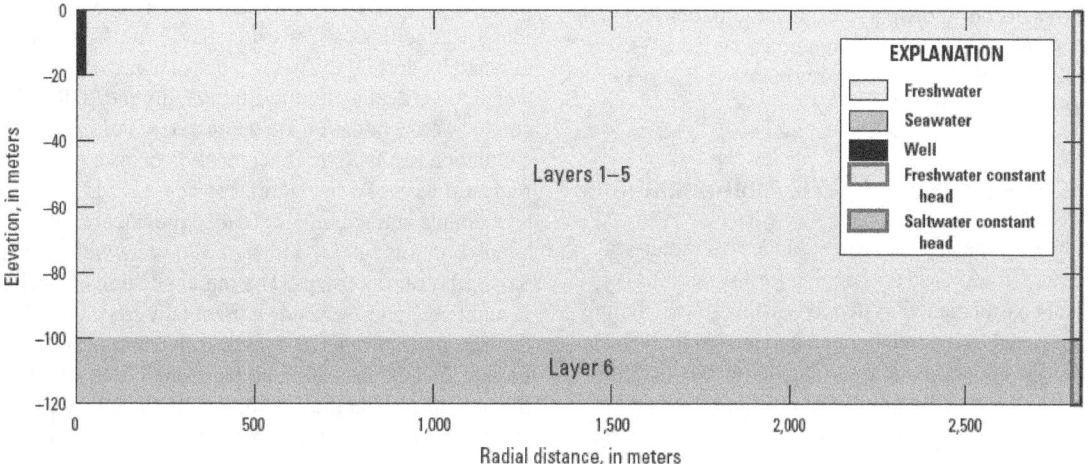

Figure 10.   The conceptual model used in example simulation 5.

all other cells, the SWI2 ISOURCE parameter was set to 0, indicating boundary conditions have water that is identical to water at the top of the aquifer.

A pumping well screened from 0 to -20 m with a withdrawal rate of 2,400 m³/d is simulated for stress period 1 at the left side of the model (column 1) in the first layer. To simulate recovery, the pumping well withdrawal rate was set to 0 m³/d for stress period 2.

The simulated position of the interface is shown at 1-year increments for the withdrawal (stress period 1) and recovery (stress period 2) periods in figure 11. During the withdrawal period, the interface steadily rises from its initial elevation of -100 m to a maximum elevation of -87 m after 4 years (fig. 11*A–D*). During the recovery period, the interface elevation decreases to -96 m at the end of year 5 (fig. 11*E*) but does not return to the initial elevation of -100 m at the end of year 8 (fig. 11*H*).

For this problem, SWI2 results are compared in figure 11 to those obtained using SEAWAT (Langevin and others, 2008). For SEAWAT, the aquifer is discretized into 120 layers that are 1 m thick, which is an increase in resolution compared to the six 20-m-thick layers used in the model simulation using the SWI2 Package. The TVD option in MT3DMS, without diffusion and dispersion, is used in the SEAWAT model. Comparison of SWI2 and SEAWAT results shows that SWI2 results are identical to the SEAWAT 50-percent seawater salinity contour at all times for horizontal distances exceeding 100 m during both the withdrawal and recovery phases of the simulation. At horizontal distances less than 100 m, the SWI2 interface overestimates the elevation of the 50-percent seawater salinity contour from the end of the second year until the end of the simulation; the discrepancy between SWI2 and SEAWAT results decreases with increasing distance from the well. The downward deflection of the interface observed during the recovery period is less for SWI2 than SEAWAT near the well. Use of additional fluid density zones has been evaluated and can be used to reduce the discrepancies between SWI2 and SEAWAT (results not shown) and evaluate salinities less than those of seawater. For example 5, run times for the SWI2 and SEAWAT simulations were approximately 3.0 and 390 seconds, respectively.

Example 6: Evaluation of SWI2 Approximations

Example 6 simulates movement of the freshwater-seawater interface in response to changing freshwater inflow in a conceptual coastal aquifer system consisting of two distinct aquifers separated by a confining unit and is used to evaluate the effect of the approximations used to derive the vertically integrated variable-density flow equations implemented in the SWI2 Package. The analyses evaluated in example 6 are based on analyses made by Dausman and others (2006) using the original SWI Package. The basic approximations used in the SWI2 Package are summarized earlier herein. These example

problems also use an arbitrary datum of 50 m instead of a sea-level datum of 0 m.

Several studies have investigated the effect of using the Dupuit approximation to simulate variable-density groundwater flow. Seawater intrusion models based on the Dupuit approximation yield accurate results for many practical problems of interface flow (for example, Bear and Dagan, 1964), even if the slope of the interface is 45° (Chan Hong and Van Duijn, 1989). Strack and Bakker (1995) showed that adoption of the Dupuit approximation for variable-density flow yields accurate results for the instantaneous flow field. Bakker and others (2004) also showed that the Dupuit approximation for a rotating brackish zone is comparable to fully three-dimensional numerical solutions.

For this example, SWI2 results are compared to results obtained with SEAWAT (Langevin and others, 2008), which simulates fully coupled variable-density groundwater flow and solute transport, including dispersion, diffusion, density inversion, and vertical resistance to flow. For SEAWAT to simulate variable-density flow and transport accurately, however, aquifers have to be discretized more finely than necessary with a MODFLOW model using the SWI2 Package, particularly in the vertical direction. Increased discretization is often required in SEAWAT (and other three-dimensional numerical solutions of variable-density groundwater flow and transport) to minimize numerical dispersion near the freshwater-seawater interface and to represent convective flow patterns. This increase results in longer model run times, which may be prohibitive for many regional-scale saltwater intrusion models. Previous evaluations of discrepancies between SWI2 and SEAWAT indicate that they produce similar results, provided that the SEAWAT model is finely discretized and numerical dispersion is minimized (for example, Bakker and others, 2004).

Problem Setup

To evaluate the appropriate usage of SWI2, a number of SWI2 and SEAWAT simulation results are evaluated to identify conditions under which SWI2 results are adversely affected by density inversions, dispersion, and increasing horizontal to vertical hydraulic conductivity ratios. Results from a single SWI2 simulation are compared with multiple SEAWAT simulation results. The conceptual model is a two-dimensional representation of a two-aquifer system (fig. 12) and is based on example simulation 3 described previously. The SWI2 simulation was run for a total of 500 years with 1 stress period having 1-year time steps. The initial position of the interface in aquifers 1 and 2 extends horizontally from 0 to 470 m along the top, slopes from 470 m at the top down to 910 m at the bottom, and extends horizontally from 910 to 4,000 m along the bottom. Within the confining unit, the interface extends from 0 to 470 m along the top and from 470 to 4,000 m along the bottom (fig. 12). All other SWI2 model parameters are identical to those specified in example simulation 3, except for the vertical datum used.

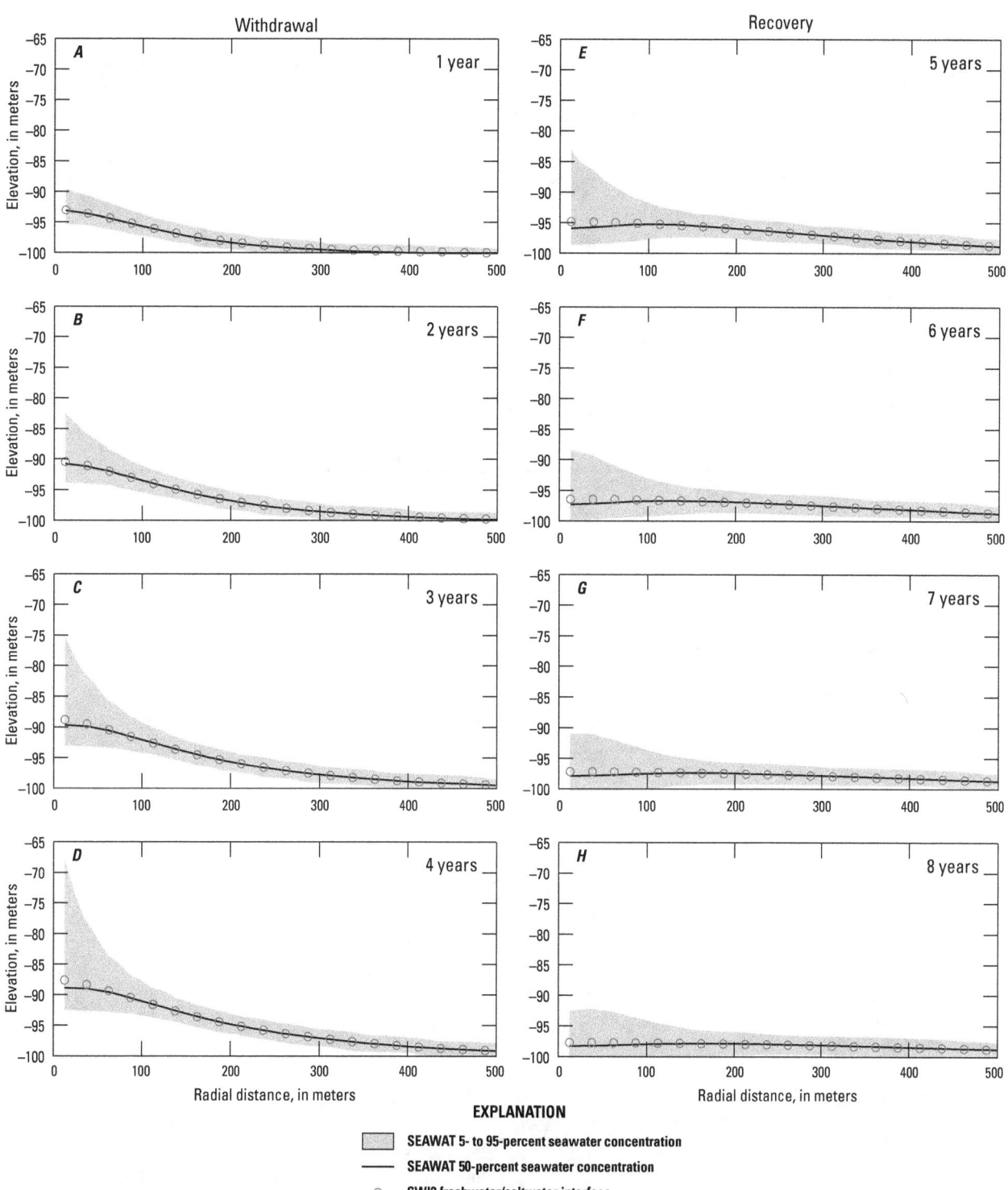

Figure 11.   Simulated freshwater-seawater interface elevation for example simulation 5 after *A,* 1; *B,* 2; *C,* 3; *D,* 4; *E,* 5; *F,* 6; *G,* 7; and *H,* 8 years. The simulated SWI2 freshwater-seawater interface is compared to simulated SEAWAT 50-percent seawater concentrations. Areas having SEAWAT simulated percent seawater concentrations ranging from 5- to 95-percent seawater are a measure of numerical dispersion.

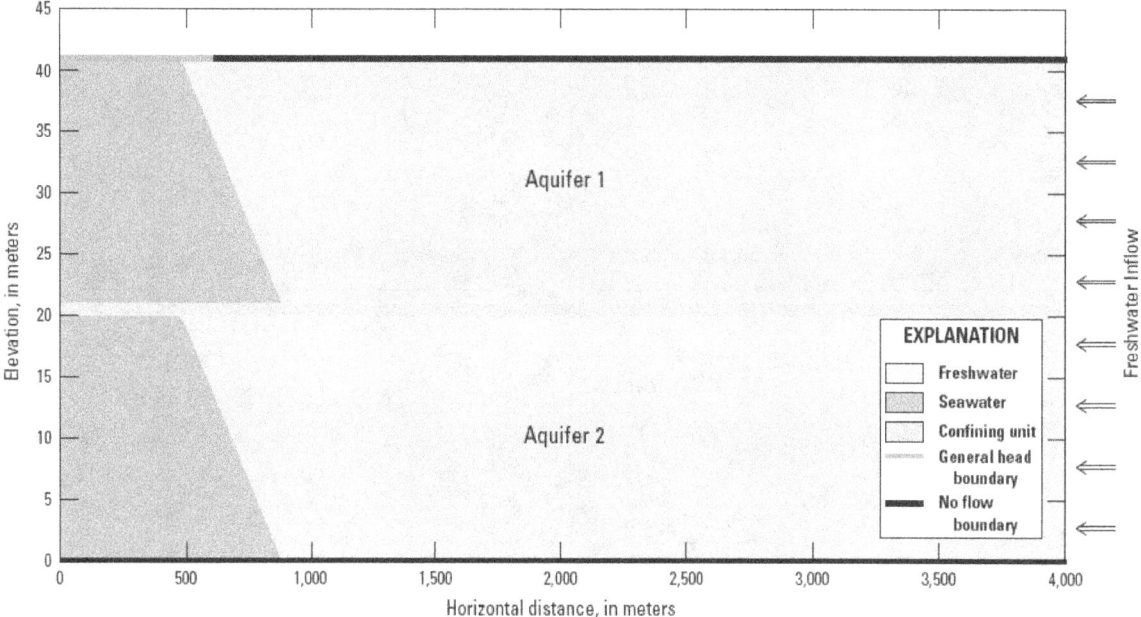

Figure 12. Conceptual model used in example simulation 6.

The SWI2 model was discretized into 200 columns that are each 20 m long (DELR), 1 row that is 1 m wide (DELC), and 3 layers that are 20, 1, and 20 m thick. For the SEAWAT simulation, the model was discretized into 800 columns that are each 5 m long (DELR), 1 row that is 1 m wide (DELC), and 82 layers that are each 0.5 m thick. The SEAWAT simulation was run for a total of 500 years and used a constant transport time step of 36.5 days. Transport was solved using the finite-difference method with upstream weighting.

For all three tests, the initial conditions represent a density inversion in which saltwater in aquifer 1 overlies freshwater in aquifer 2 over part of the simulation domain (fig. 12). In the first test, which evaluates density inversion (E6-1), SEAWAT dispersivity values were set to zero and the horizontal to vertical hydraulic conductivity ratio was set to 1. The second test (E6-2) evaluates increasing levels of dispersion by comparing SEAWAT and SWI2 simulation results. For example simulation E6-2, the transverse dispersivity was specified to be an order of magnitude less than the longitudinal value. The third test evaluates decreasing vertical hydraulic conductivity of aquifers 1 and 2 (E6-3) by comparing SEAWAT and SWI2 simulation results.

Density Inversion (E6-1)

At 100 years, the SWI2 and SEAWAT simulation results agree closely, except near the density inversion as shown in figure 13. For the SWI2 simulation, seawater migrating vertically from aquifer 1 is added to the seawater zone in aquifer 2. Although dispersivity is set to zero, a transition zone is simulated as a result of numerical dispersion in SEAWAT. The final location of the interface (fig. 13B) at 500 years simulated by SWI2 agrees closely with the 50-percent seawater salinity contour simulated with SEAWAT. However, slight differences observed include (1) the position of the interface at the bottom of aquifer 1 and top of aquifer 2 is discontinuous for the SWI2 simulation only, and (2) the toe of the interface for the SEAWAT model is slightly seaward of the SWI2 surface as a result of dispersion and mixing of freshwater moving vertically between the two aquifers. For this test, run times for the SWI2 and SEAWAT simulations were approximately 0.1 and 4.0 seconds, respectively.

Dispersion (E6-2)

Model results indicate that as dispersivity increases, the location of the toe moves seaward in both aquifers (fig. 14A). Results indicate that SWI2 produces reasonable results if the width of the transition zone from freshwater to seawater is less than 18 percent of the distance between the position of the interface at the top and bottom of the aquifer (fig. 14B). In this case, the transition zone is the distance between the 5- and 95-percent seawater salinity contours at the base of the aquifer, and the interface is the 50-percent seawater salinity contour. For this problem, longitudinal dispersivity values greater than a few meters result in notable differences between SWI2 and SEAWAT. Typically, calibrated values of the dispersivity for many seawater intrusion models having similar cell sizes are less than a couple of meters (for example, Oude Essink, 2001); therefore, SWI2 is probably appropriate for many saltwater

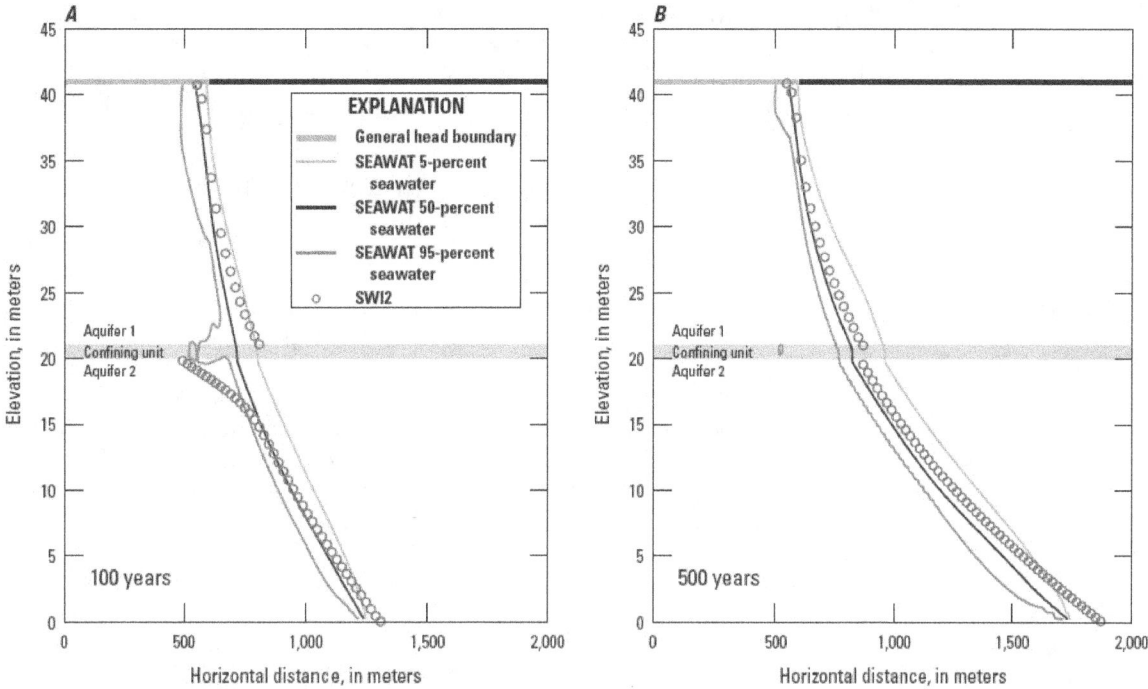

Figure 13. Comparison of SWI2 (open blue circles) and SEAWAT interface positions after *A,* 100 and *B,* 500 years for the case where a density inversion exists. The SEAWAT interface is defined as the position of the 50-percent seawater concentration contour. Note the discontinuity of the SWI2 freshwater-seawater interface at the bottom of aquifer 1 and top of aquifer 2 and that the toe of the SWI2 interface is landward of the SEAWAT freshwater-seawater interface at the bottom of aquifer 1.

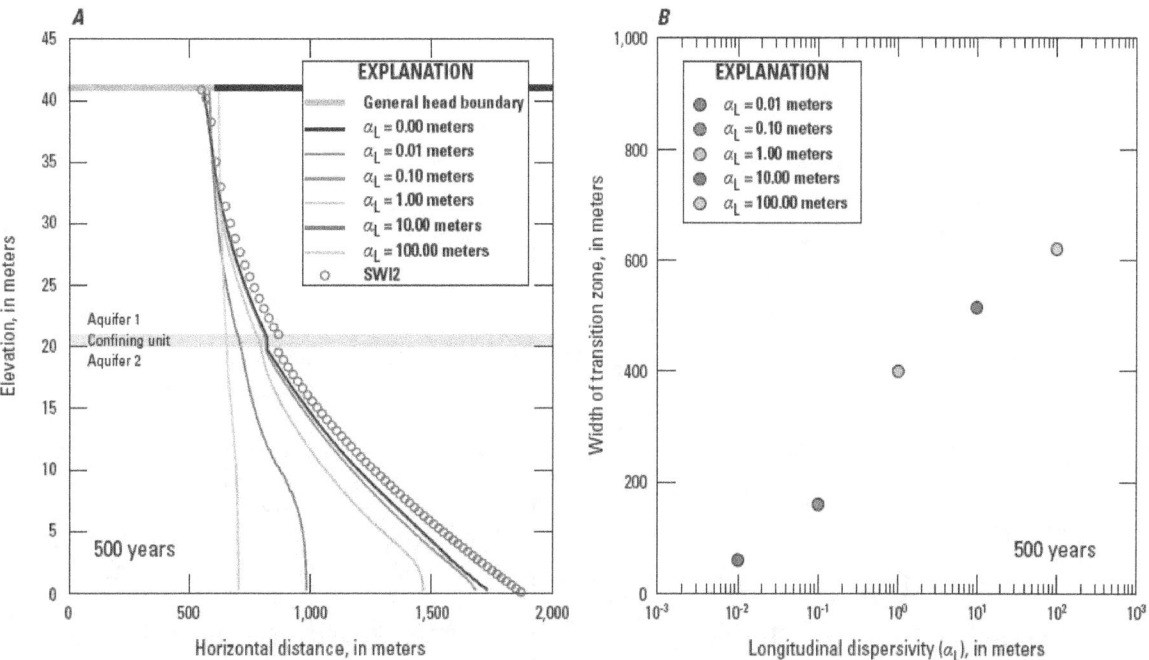

Figure 14. *A,* comparison of the SWI2 freshwater-seawater interface (open blue circles) and the SEAWAT 50-percent seawater concentration contour (solid lines) simulated using longitudinal dispersivity (α_L) values of 0.0, 0.01, 0.1, 1, 10, and 100 meters after 500 years; and *B,* the relation of longitudinal dispersivity (α_L) to the width of the transition zone, simulated using SEAWAT and defined as the distance between the 5- and 95-percent seawater concentrations at the base of aquifer 2 (model layer 3), after 500 years.

intrusion models. For problems having different cell sizes, acceptable longitudinal dispersivities will probably scale with cell size, and SWI2 users are encouraged to determine the validity of using SWI2 by using synthetic test problems having comparable dimensions and boundary conditions. For this test, the run time for the SWI2 simulation was approximately 0.4 second and the run times for the SEAWAT simulations ranged from approximately 1,000 to 1,600 seconds.

Vertical Resistance To Flow (E6-3)

As vertical resistance to flow is increased in aquifers 1 and 2, the tip and toe of the interface move seaward. For the SWI2 simulations, this is the case when the horizontal to vertical hydraulic conductivity ratio ($K_h : K_v$) is greater than 10 (figure 15A). For the SEAWAT simulations, the tip and toe of the interface move seaward when $K_h : K_v$ exceeds 10 (figure 15A). Comparison of results indicates the difference between the toe position of SWI2 and SEAWAT simulations remains approximately constant (average difference = 155 m) with an increasing $K_h : K_v$ ratio (figure 15B). Conversely, the comparison indicates the difference between the tip position of SWI2 and SEAWAT simulations is negatively correlated with an increasing $K_h : K_v$ ratio (figure 15B). The difference between SWI2 and SEAWAT tip positions increases as vertical head gradients within each aquifer increase and the Dupuit

approximation used to formulate SWI2 becomes less appropriate for a given problem. For this test, run times for the SWI2 and SEAWAT simulations ranged from approximately 0.4 to 0.6 second and 1,200 to 1,300 seconds, respectively.

Example 7: Regional Problem

Example 7 represents an application of the SWI2 Package in a regional-scale model to simulate the steady-state position of the freshwater-seawater interface in the shallow, unconfined aquifer underlying Cape Cod, Massachusetts (fig. 16). The aquifer is the sole source of water to local communities (Walter and Whealan, 2005). The Cape Cod model was developed from the models described in Masterson (2004), Walter and Whealan (2005), and Masterson and others (2009). The base of the aquifer was delineated using updated bedrock topography data from Fairchild and others (2013). Readers are referred to these documents for further information about the aquifer and (or) the development and calibration of the MODFLOW models developed as part of these studies. The steady-state position of the freshwater-seawater interface and groundwater heads in this example model were not calibrated and is intended to be a demonstration of the application of the SWI2 Package for realistic conditions.

The domain is discretized into 2,272 columns and 1,384 rows with a constant 30.48-m grid spacing (DELR and DELC)

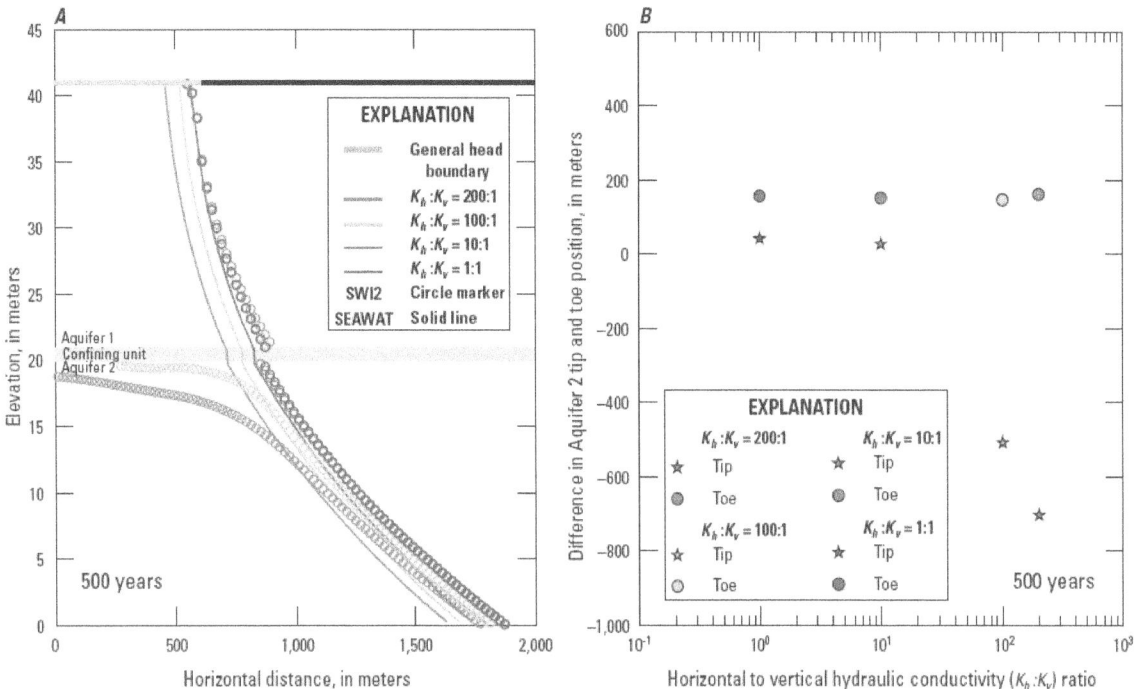

Figure 15. *A*, comparison of the SWI2 freshwater-seawater interface (open circles) and SEAWAT 50-percent seawater concentration contour (solid lines) after 500 years for simulations with horizontal to vertical hydraulic conductivity ratios (K_h:K_v) of 1:1, 10:1, 100:1, and 200:1; and *B*, the relation of horizontal to vertical hydraulic conductivity ratios (K_h:K_v) to differences between the SWI2 and SEAWAT calculated tip and toe positions in aquifer 2 after 500 years.

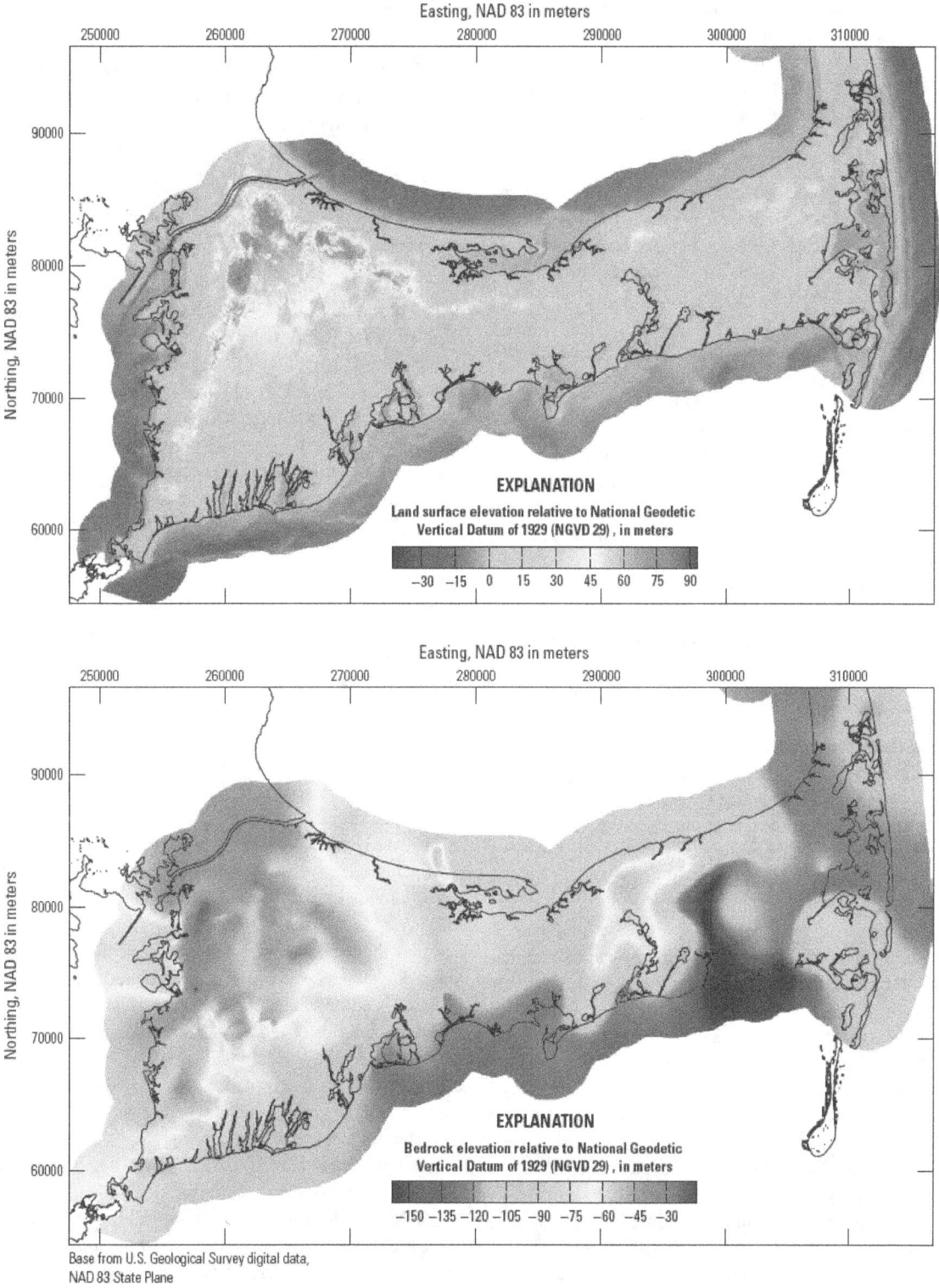

Figure 16. *A*, land-surface and *B*, bedrock elevations in example simulation 7.

and 1 layer with a variable surface and aquifer bottom elevation, as shown in figure 16. The active model area is indicated
by the colored area shown in figure 16 and includes a total
of 1,699,730 model cells. A single steady-state MODFLOW
stress period was simulated. A number of time steps were run in
order to allow the interface to equilibrate with model boundary
conditions.

The hydraulic conductivity of the aquifer is heterogeneous
and is shown in figure 17A. Hydraulic conductivities range
from 0.3 to 30,000 m/d, and are based on data from Masterson
(2004), Walter and Whealan (2005), and Masterson and others
(2009). A constant effective porosity of 0.2 was specified for the
aquifer.

Offshore, general head boundary conditions were used to
represent the ocean boundary. The offshore bathymetry was
used to calculate freshwater heads at the bottom of the ocean
(or top of the aquifer) for the coastal general head boundaries.
All other general head boundary condition data and all data for
drain (DRN) boundary conditions were derived from Masterson
(2004), Walter and Whealan (2005), and Masterson and others
(2009). Drain boundaries were used to represent freshwater
lakes and surface-water features in the model.

The groundwater is divided into a freshwater zone and a
seawater zone, separated by an active *ZETA* surface, ζ_2, between
the zones (NSRF=1) that approximates the 50-percent seawater
salinity contour. Fluid density is represented using the stratified option (ISTRAT=1). The dimensionless density difference between freshwater and saltwater is 0.025. The tip and
toe tracking parameters are a TOESLOPE and TIPSLOPE of
0.025, a default ALPHA of 0.1, and a default BETA of 0.1. The
initial freshwater-seawater interface was calculated using initial
heads and the Ghyben-Herzberg relation. The SWI2 ISOURCE
parameter is set to -2 in all general head boundaries representing
coastal boundaries, which ensures that inflow from the coastal
boundaries is saltwater and outflow is from the top zone, which
can be freshwater. In all other cells, the SWI2 ISOURCE parameter was set to 0, indicating boundary conditions have water that
is identical to water at the top of the aquifer.

A spatially distributed areal recharge rate averaging 686
millimeters per year (mm/yr), was applied for onshore portions
of the model and is shown in figure 17B. A total of 191 groundwater wells, having a total withdrawal rate of 96,774 m³/d, were
simulated in the model. The areal recharge and groundwater
withdrawal rates are based on data from Masterson (2004), Walter and Whealan (2005), and Masterson and others (2009).

Simulated steady-state groundwater levels for the model
are shown in figure 18. The simulated distribution of freshwater and saltwater in three north-south cross sections is shown
in figure 19. The cross sections indicate that the freshwater
lens extends to the base of the aquifer over much of the extent
portrayed. The freshwater lens is thickest in the eastern part of
the aquifer and areally corresponds to the area having the lowest
bedrock elevation (fig. 20A). The simulated elevation of the
interface is shown in figure 20B, as are areas where the freshwater lens extends to the base of the aquifer.

Summary

The SWI2 Package for MODFLOW is capable of simulating vertically integrated variable-density groundwater flow
in layered aquifer systems. The formulation used in the SWI2
Package is based on the Dupuit approximation and requires
discretization of groundwater flow within aquifers into zones
of varying density. Use of the Dupuit approximation allows
vertically integrated variable-density groundwater flow in indi-
vidual aquifers to be represented using a single layer of cells,
which greatly reduces vertical discretization and the number of
cells required to accurately simulate interface movement and
regional groundwater flow. The resulting differential equations
are similar in form to the groundwater flow equation solved
by MODFLOW and only require (1) the addition of pseudo-
source terms to reformulate the single-density groundwater
flow equation to solve for vertically integrated variable-density groundwater flow, and (2) separate solutions for interface
movement using flux rates calculated by the groundwater
flow equation. This approach allows the SWI2 Package to be
implemented as a standard MODFLOW package that calcu-
lates and adds terms to the groundwater flow equation solved
by MODFLOW.

Use of the SWI2 Package in MODFLOW only requires
the addition of a single additional input file and specification
of boundary heads as freshwater heads at the top of the aquifer. The fluid density distribution within groundwater model
layers can be represented using zones of constant density
(stratified flow) or continuously varying density (piecewise
linear in the vertical direction) in the SWI2 Package. The
SWI2 Package also includes options for using (1) direct (DE4)
or iterative (PCG) solvers to solve for interface movement and
(2) smaller adaptive SWI2 time steps within MODFLOW time
steps to refine interface movement and increase numerical
stability.

For the example problems evaluated, the position of
the freshwater-seawater interface simulated using the SWI2
Package is comparable to the interface positions obtained from
analytical solutions or SEAWAT simulations. SWI2 results
may not be realistic when the dispersion across an interface
is large, when the horizontal to vertical hydraulic conductivity ratio is large, or in systems where inversion occurs and
a substantial amount of vertical fingering is observed. For
simulations in which SWI2 and SEAWAT results diverge, the
SWI2-simulated interface tends to be more landward than the
SEAWAT-simulated transition zone. In terms of evaluating the
risk of saltwater intrusion to freshwater resources, the SWI2
results could be considered more conservative. The com-
putational savings obtained using SWI2 instead of coupled
variable-density groundwater flow and transport codes such
as SEAWAT are substantial, decreasing run times from a few
hours to a few seconds. These savings make the SWI2 Package a valid, time-saving alternative for many regional-scale
groundwater models.

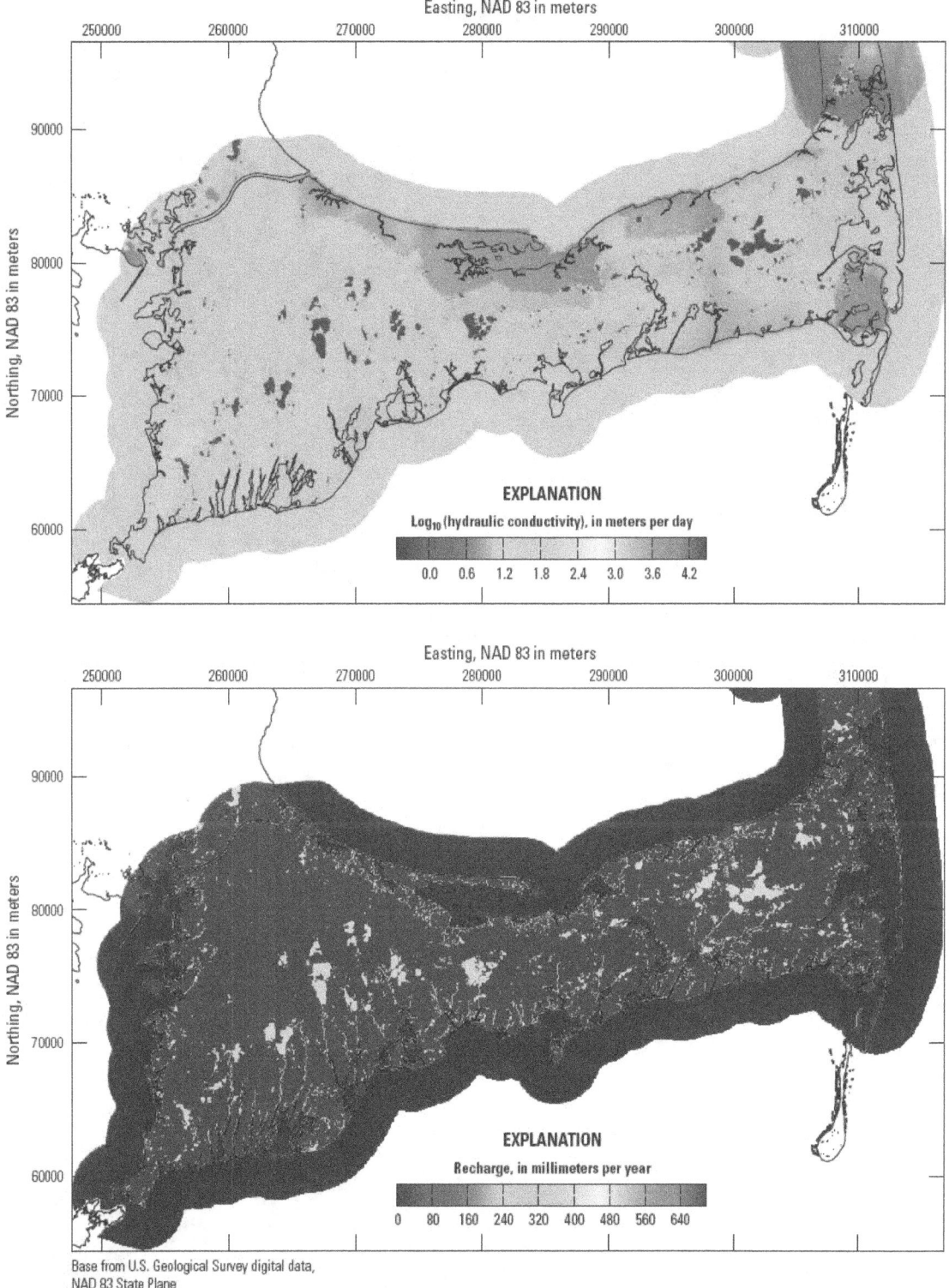

Figure 17.   *A*, hydraulic conductivity and *B*, areal recharge in example simulation 7.

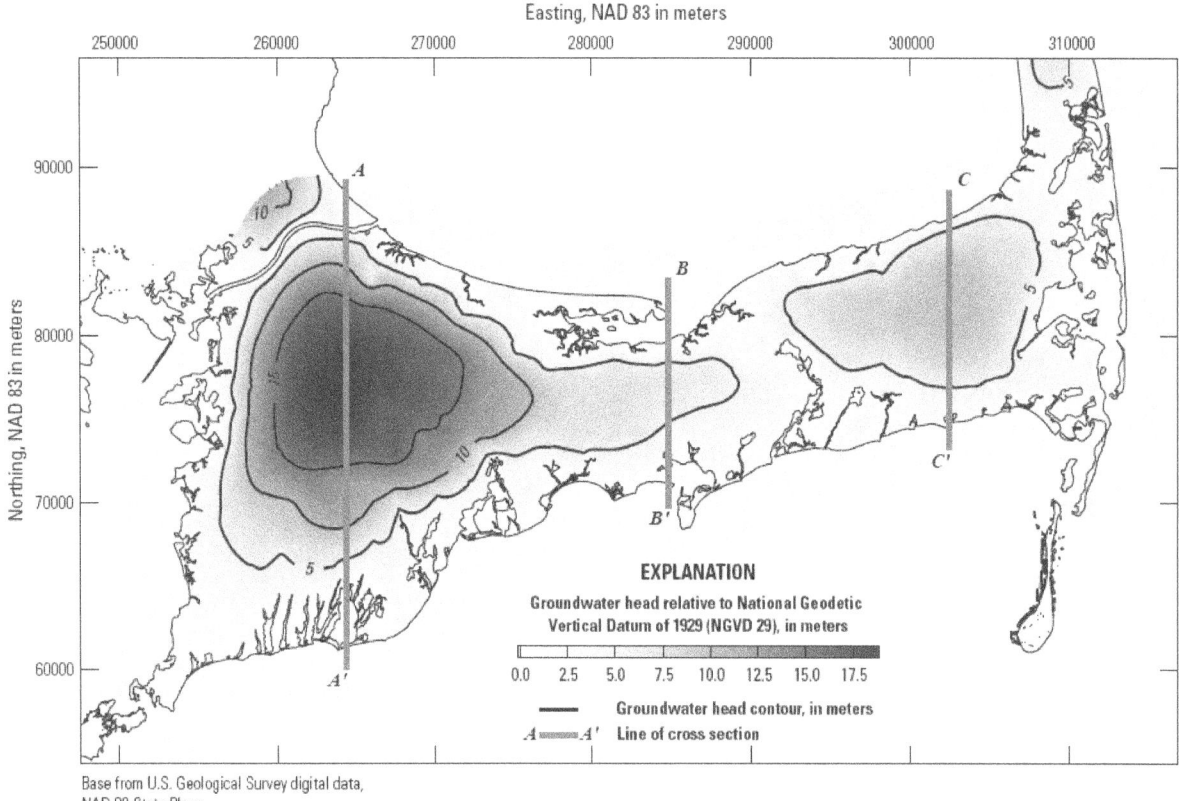

Figure 18.   Simulated groundwater heads and the location of cross sections *A–A′, B–B′,* and *C–C′* for example simulation 7.

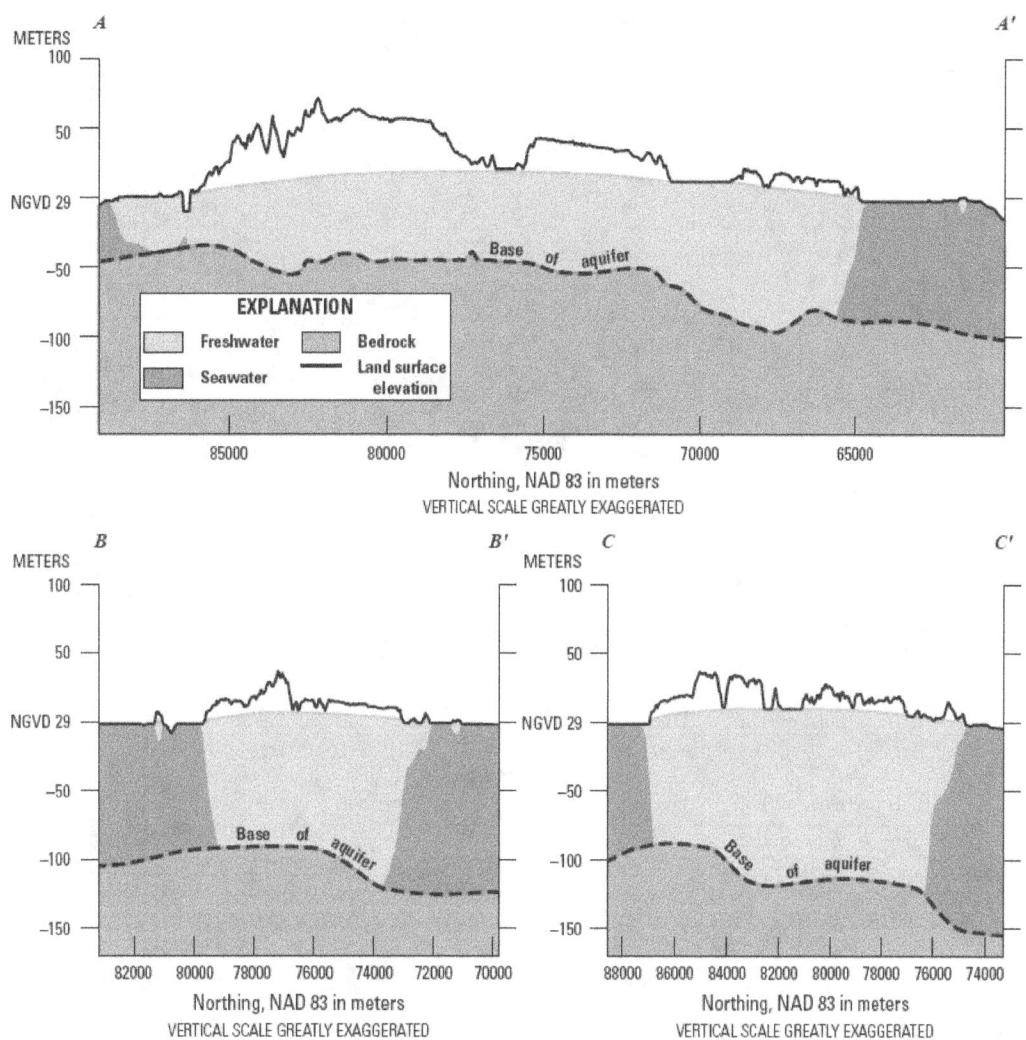

Figure 19. Simulated distribution of freshwater and seawater along cross sections *A–A′*, *B–B′*, and *C–C′* for example simulation 7.

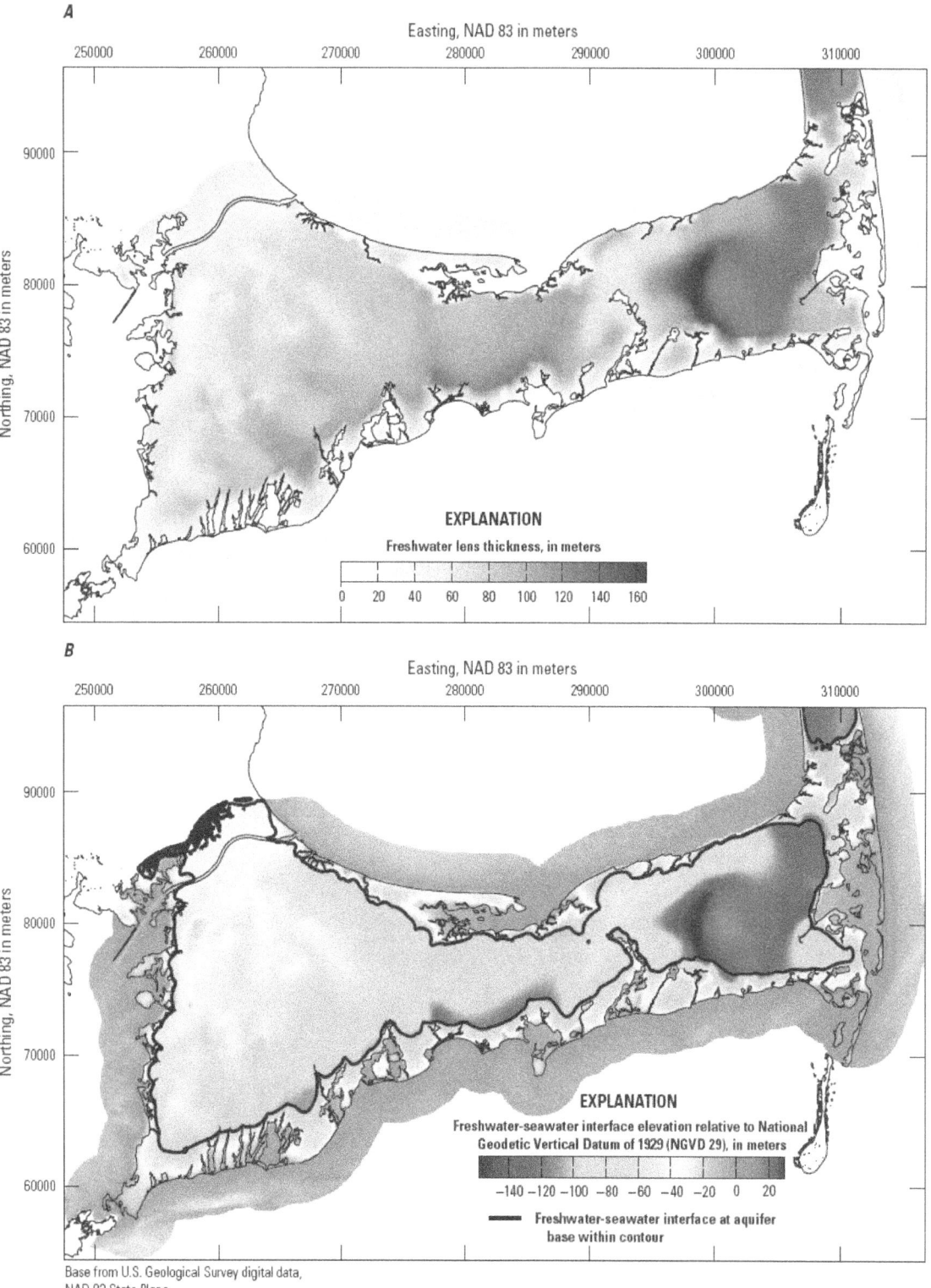

Base from U.S. Geological Survey digital data,
NAD 83 State Plane

Figure 20. Simulated *A*, freshwater lens thickness and *B*, the freshwater-seawater interface elevation for example simulation 7. The black contour lines in *B* encompass areas where the freshwater lens extends to the base of the aquifer.

References Cited

Anderson, M.P., and Woessner, W.W., 1992, Applied ground-water modeling: Simulation of flow and advective transport: San Diego, Calif., Academic Press, 381 p.

Bakker, Mark, 2003, A Dupuit formulation for modeling seawater intrusion in regional aquifer systems: Water Resources Research, v. 39, no. 5, p. 1131–1140.

Bakker, Mark, Oude Essink, G.H.P., and Langevin, C.D., 2004, The rotating movement of three immiscible fluids—A benchmark problem: Journal of Hydrology, v. 287, p. 271–279.

Bakker, Mark, and Schaars, Frans, 2011, The Sea Water Intrusion (SWI) Package manual: Part I. Theory, user manual, and examples (Version 1.2): Available at modflowswi.googlecode.com.

Bear, Jacob, and Dagan, Gedeon, 1964, Some exact solutions of interface problems by means of the hodograph method: Journal of Geophysical Research, v. 69, no. 8, p. 1563–1572.

Chan Hong, J.R., and van Duijn, C.J., 1989, The interface between fresh and salt groundwater: A numerical study: IMA Journal of Applied Mathematics, v. 42, p. 209–240.

Dausman, A.M., Langevin, C.D., Bakker, Mark, and Schaars, Frans, 2010, A comparison between SWI and SEAWAT—The importance of dispersion, inversion and vertical anisotropy: in Condesso de Melo, M.T., Lebbe, L., Cruz, J.V., Coutinho, Rui, Langevin, C.D., and Buxo, Ana eds., Proceedings of the 21st Salt Water Intrusion Meeting, June 21–26, 2010, Ponta Delgada, São Miguel, Azores, Portugal.

Diersch, H.J.G., and Kolditz, Olaf, 2002, Variable-density flow and transport in porous media—Approaches and challenges: White Papers vol. II, accessed June 3, 2012, at http://www.feflow.info/manuals.html.

Essaid, H.I., 1990, SHARP—A quasi-three-dimensional finite-difference simulation model for freshwater and saltwater flow in layered coastal aquifer systems: U.S. Geological Survey Water-Resources Investigations Report 90–4130, 181 p.

Faichild, G.M., Lane, J.W. Jr., Voytek, E.B., and LeBlanc, D.R., 2013, Bedrock topography of western Cape Cod, Massachusetts, based on bedrock altitudes from geologic borings and ambient seismic noise by the horizontal-to-vertical spectral-ratio method: U.S. Geological Survey Scientific Investigations Map 3233 [one plate and accompanying brochure].

Ghyben, W.B., 1889, Nota in verband met de voorgenomen putboring nabij Amsterdam: Tijdschrift van het Koninklijk Instituut van Ingenieurs, p. 8–22.

Harbaugh, A.W., 2005, MODFLOW–2005, the U.S. Geological Survey modular ground-water model—The ground-water flow process: U.S. Geological Survey Techniques and Methods, book 6, chap. A16.

Harbaugh, A.W., Banta, E.R., Hill, M.C., and McDonald, M.G., 2000, MODFLOW–2000, the U.S. Geological Survey modular ground-water model—User guide to modularization concepts and the Ground-Water Flow Process: U.S. Geological Survey Open-File Report 00–92, 121 p.

Harbaugh, A.W., and McDonald, M.G., 1996, User's documentation for MODFLOW-96—An update to the U.S. Geological Survey modular finite-difference ground-water flow model: U.S. Geological Survey Open-File Report 96–485, 56 p.

Herzberg, A., 1901, Die Wasserversorgung einiger Nordsee-bäder: Journal für Gasbeleuchtung und Wasserversorgung, v. 44, p. 815–819.

Holzbecher, E.O., 1998, Modeling density-driven flow in porous media—Principles, numerics, software: Berlin, Springer-Verlag, 286 p.

Langevin, C.D., 2008, Modeling axisymmetric flow and transport: Ground Water, v. 46, p. 579–590, (Available online at http://dx.doi.org/10.1111/j.1745-6584.2008.00445.x)

Langevin, C.D., Thorne, D.T., Jr., Dausman, A.M., Sukop, M.C., and Guo, Weixing, 2008, SEAWAT Version 4—A computer program for simulation of multi-species solute and heat transport: U.S. Geological Survey Techniques and Methods, book 6, chap. A22, 39 p.

Maas, C., and Emke, M.J, 1988, Solving varying density groundwater problems with a single density computer program: Natuurwetenschappelijk Tijdschrift, v. 70, p. 143–154.

Masterson, J.P., 2004, Simulated interaction between freshwater and saltwater and effects of ground-water pumping and sea-level change, Lower Cape Cod aquifer system, Massachusetts: U.S. Geological Survey Scientific Investigations Report 2004–5014, 72 p.

Masterson, J.P., Carlson, C.S., and Walter, D.A., 2009, Hydrogeology and simulation of groundwater flow in the Plymouth-Carver-Kingston-Duxbury aquifer system, southeastern Massachusetts: U.S. Geological Survey Scientific Investigations Report 2009–5063, 110 p.

McDonald, M.G., and Harbaugh, A.W., 1988, A modular three-dimensional finite-difference ground-water flow model: U.S. Geological Survey Techniques of Water-Resources Investigations, book 6, chap. A1, 586 p.

Oude Essink, G.H.P., 2001, Salt water intrusion in a three-dimensional groundwater system in the Netherlands—A numerical study: Transport in Porous Media, v. 43, no. 1, p. 137–158.

Post, Vincent, Kooi, Henk, and Simmons, Craig, 2007, Using hydraulic head measurements in variable-density ground water flow analyses: Ground Water, v. 45, p. 664–671 (Available online at http://dx.doi.org/10.1111/j.1745-6584.2007.00339.x).

Strack, O.D.L., 1989, Groundwater mechanics: Englewood Cliffs, N.J., Prentice Hall.

Strack, O.D.L., 1995, A Dupuit-Forchheimer model for three-dimensional flow with variable density: Water Resources Research, v. 31, no. 12, p. 3007–3017.

Strack, O.D.L., and Bakker, Mark, 1995, A validation of a Dupuit-Forchheimer formulation for flow with variable density: Water Resources Research, v. 31, no. 12, p. 3019–3024.

Taylor, J.Z., and Person, Mark, 1998, Capture zone delineations on island aquifer systems: Ground Water, v. 36, p. 722–730 (Available online at http://dx.doi.org/10.1111/j.1745-6584.1998.tb02189.x).

Voss, C.I., and Provost, A., 2010, SUTRA—A model for saturated-unsaturated variable-density ground-water flow with solute or energy transport: U.S. Geological Survey Water-Resources Investigations Report 02–4231 (Version of September 22, 2010 (SUTRA Version 2.2).

Walter, D.A., and Whealan, A.T., 2005, Simulated water sources and effects of pumping on surface and ground water, Sagamore and Monomoy flow lenses, Cape Cod, Massachusetts: U.S. Geological Survey Scientific Investigations Report 2004–5181, 85 p

Weiss, Emanuel, 1982, A model for the simulation of flow of variable-density ground water in three dimensions under steady-state conditions: U.S. Geological Survey Open File Report 82–352, 59 p.

Wilson, J.L., and Sa Da Costa, Antonio, 1982, Finite element simulation of a saltwater/freshwater interface with indirect toe tracking: Water Resources Research, v. 18, no. 4, p. 1069–1080.

Wooding, R.A., 1969, Growth of fingers at an unstable diffusing interface in a porous medium or Hele-Shaw cell: Journal of Fluid Mechanics, v. 39, no. 3, p. 477–495.

Zhou, Q., Bear, J., and Bensabat, J., 2005, Saltwater —Upconing and decay beneath a well pumping above an interface zone: Transport in Porous Media, v. 61, no. 3, p. 337–363.

Appendixes 1 and 2

Appendix 1.   Input Instructions

The use of the Seawater Intrusion (SWI2) Package is an advanced application of MODFLOW, and it is assumed that users are familiar with the use of MODFLOW and the input files required for MODFLOW as documented in Harbaugh (2005); thus, this appendix only describes input files required by SWI2.

MODFLOW Name (NAM) File

Simulation of vertically integrated variable-density groundwater flow using the seawater intrusion (SWI2) Package is activated by including a record in the MODFLOW name file using the file type (Ftype) "SWI2" to indicate that relevant calculations are to be made in the model and to specify the related input data file. The NAM file should also be modified to include the appropriate information for saving interface (ZETA) elevations, SWI2 cell-by-cell budget data, and (or) SWI2 observation well data, if necessary.

Basic (BAS) Package Input Instructions

Starting heads (STRT) should be modified to represent the freshwater head at the top of each layer. At a minimum, starting heads should be modified for all constant head cells representing fluid densities greater than freshwater. Converting all starting heads to freshwater heads may be advantageous for transient simulations in which good starting heads may improve simulated heads. Freshwater heads can be calculated using equation 3.

Output (OC) Control Input Instructions

If desired, ZETA output frequency is controlled by SAVE HEAD or Hdsv variables in the output control file. Similarly, output of SWI2 Package cell-by-cell data is controlled by SAVE BUDGET or ICBCFL variables in the output control file.

Head-Dependent Boundary Conditions Input Instructions

All standard MODFLOW packages may be used with the SWI2 Package. When the package requires specification of a head, for example the General Head Boundary (GHB) Condition Package, this head *must* be the freshwater head at the top of the layer. Modification of head-dependent boundary-condition specified-heads or threshold elevations is only required for boundary conditions representing fluids having densities other than that of freshwater or in areas where simulated freshwater heads reflect groundwater that is brackish to saline. For example, in a model setup using general head boundaries, heads for coastal boundaries should be specified as freshwater heads, and heads for general head boundaries representing inland, fresh sources of water can be left unmodified. Boundary conditions that may need modification include River (RIV), Drain (DRN), Drain Return (DRT), General Head Boundary (GHB), Streamflow Routing (SFR), Lake (LAK) Surface-Water Routing (SWR), Evapotranspiration (EVT), and Evapotranspiration Segments (ETS) Packages. Freshwater heads can be calculated using equation 3.

Time-Variant Specified-Head (CHD) Package Input Instructions

Similar to constant- and initial-head (STRT) values in the Basic (BAS) Package, time-variant specified head values for cells representing fluid densities greater than freshwater should be modified to represent the freshwater head at the top of each cell. Freshwater heads can be calculated using equation 3.

SWI2 Data Input Instructions

The SWI2 file contains solver variables, data values for the different zones, algorithm parameters, the initial positions of the surfaces, the type of sources and sinks, and *ZETA* observation locations. In the body of the report, the number of surfaces are defined as being equal to the number of zones plus one. This number includes overlying and underlying surfaces corresponding to the layer top and bottom. In the input and output described here, however, only active *ZETA* surfaces are read and written. Active surfaces include only those surfaces between zones, and do not include the surfaces corresponding to the layer top and bottom. Thus, the number of active surfaces is equal to the number of zones minus one. Optional variables are indicated in [brackets].

FOR EACH SIMULATION
1.   Data: NSRF ISTRAT NOBS ISWIZT ISWIBD ISWIOBS [OPTIONS]
        Module: URWORD

2a.   Data: NSOLVER IPRSOL MUTSOL
        Module: URWORD

IF NSOLVER = 2
2b.   Data: MXITER ITER1 NPCOND ZCLOSE RCLOSE RELAX NBPOL DAMP [DAMPT]
        Module: URWORD

3a.   Data: TOESLOPE TIPSLOPE [ALPHA] [BETA]
        Module: URWORD

IF OPTIONS = ADAPTIVE
3b.   Data: NADPTMX NADPTMN ADPTFCT
        Module: URWORD

4.   Data: NU(ISTRAT=0: NSRF+2, ISTRAT=1: NSRF+1)
        Module: U1DREL

FOR EACH SURFACE (FROM 1 to NSRF)
  FOR EACH LAYER
5.   Data: ZETA(NCOL,NROW)
        Module: U2DREL

FOR EACH LAYER
6.   Data: SSZ(NCOL,NROW)
        Module: U2DREL

FOR EACH LAYER
7.   Data: ISOURCE(NCOL,NROW)
        Module: U2DREL

IF NOBS > 0
  FOR EACH NOBS
8.   Data: OBSNAM LAYER ROW COLUMN
        Module: URWORD

Explanation of variables read by the SWI2 Package

NSRF—Number of active surfaces (interfaces). This equals the number of zones minus one.

ISTRAT—Flag indicating density distribution.

0 – density varies linear between surfaces.
1 – density is constant between surfaces.

NOBS—Number of OBS observation locations.

ISWIZT—Flag and a unit number for ZETA output.

> If ISWIZT > 0, unit number for ZETA output
> If ISWIZT ≤ 0, ZETA will not be recorded.

> If ISWIZT > 0, ZETA output is written with the same frequency as HEAD output specified in the Output Control (OC) file.

ISWIBD—Flag and a unit number for BUDGET output. When this option is selected, corrections to the cell by cell flows computed by MODFLOW will be written to the same or different file (depending on the unit number). Corrections are called SWIADDTOFLF, SWIADDTOFRF, and SWIADDTOFFF, for the lower face (LF), right face (RF) and front face (FF), respectively

> If ISWIBD > 0, unit number for BUDGET
> If ISWIBD ≤ 0, BUDGET will not be recorded.

ISWIOBS—Flag and a unit number for OBS output

> If ISWIOBS > 0, unit number for ASCII OBS
> If ISWIOBS = 0, OBS will not be recorded.
> If ISWIOBS < 0, |ISWIOBS| unit number for binary OBS.

OPTIONS—Is an optional list of character values

> "ADAPTIVE" —defines that adaptive SWI2 time step variables will be specified in dataset 3b.

NSOLVER—Flag indicating solver used to solve for ZETA surfaces

> If NSOLVER = 1, the MODFLOW DE4 solver will be used.
> If NSOLVER = 2, the MODFLOW PCG solver will be used. The PCG solver should be used for large problems when the time required for the DE4 solver to obtain a solution is excessive.

IPRSOL—is the printout interval for printing convergence information. If IPRSOL is equal to zero, it is changed to 999. The maximum ZETA change (positive or negative) and residual change are printed for each iteration of a time step whenever the time step is an even multiple of IPRSOL.

MUTSOL—is a flag that controls printing of convergence information from the solver.

> If MUTSOL = 0, tables of maximum head change and residual will be printed each iteration.
> If MUTSOL = 1, only the total number of iterations will be printed.
> If MUTSOL = 2, no information will be printed.
> If MUTSOL = 3, information will only be printed if convergence fails.

MXITER—Maximum number of outer iterations—that is, calls to the solution routine.

ITER1—Maximum number of inner iterations—that is, iterations within the solution routine.

NPCOND—flag used to select the matrix conditioning method for the MODFLOW PCG solver (NSOLVER = 2).

> If NPCOND = 1, is for Modified Incomplete Cholesky (for use on scalar computers).
> If NPCOND = 2, is for Polynomial (for use on vector computers or to conserve computer memory).

ZCLOSE—is the ZETA change criterion for convergence, in units of length. When the maximum absolute value of ZETA change from all nodes during an iteration is less than or equal to ZCLOSE, and the criterion for RCLOSE is also satisfied (see below), iteration stops.

RCLOSE—is the residual criterion for convergence, in units of cubic length per time. The units for length and time are the same as established for all model data. (See LENUNI and ITMUNI input variables in the Discretization File.) When the maximum absolute value of the residual at all nodes during an iteration is less than or equal to RCLOSE, and the criterion for ZCLOSE is also satisfied (see above), iteration stops.

RELAX—is the relaxation parameter used with NPCOND = 1. Usually, RELAX = 1.0, but for some problems a value of 0.99, 0.98, or 0.97 will reduce the number of iterations required for convergence. RELAX is only used if NSOLVER is 2 and NPCOND is 1.

NBPOL—is only specified when NSOLVER = 2 and used when NPCOND = 2 to indicate whether the estimate of the upper bound on the maximum eigenvalue is 2.0, or whether the estimate will be calculated. NBPOL = 2 is used to specify the value is 2.0; for any other value of NBPOL, the estimate is calculated. Convergence is generally insensitive to this parameter. NBPOL is not used if NPCOND is not 2.

DAMP—is the damping factor and is only specified when NSOLVER = 2. It is typically set equal to one, which indicates no damping. A value less than 1.0 and greater than 0.0 causes damping.

     If DAMP > 0, applies to both steady-state and transient stress periods.
     If DAMP < 0, applies to steady-state periods. The absolute value if used as the dampening factor.

DAMPT—is the damping factor for transient stress periods and is only specified when NSOLVER = 2. DAMPT is enclosed in brackets indicating that it is an optional variable that only is read when DAMP is specified as a negative value. If DAMPT is not read, then the single damping factor, DAMP, is used for both transient and steady-state stress periods.

TOESLOPE—Maximum slope of toe cells.

TIPSLOPE—Maximum slope of tip cells.

ALPHA—Fraction of threshold used to move the tip and toe to adjacent empty cells when the slope exceeds user-specified TOESLOPE and TIPSLOPE values. If ALPHA is not specified, a default value of 0.1 is used. ALPHA must be greater than 0.0 and less than or equal to 1.0.

BETA— Fraction of threshold used to move the toe to adjacent non-empty cells when the surface is below a minimum value defined by the user-specified TOESLOPE value. A default BETA value of 0.1 is used if ALPHA is not specified. BETA must be greater than 0.0 and less than or equal to 1.0.

NADPTMX—Maximum number of SWI2 time steps per MODFLOW time step. If NADPTMX is less than 1, it is changed to 1.

NADPTMN—Minimum number of SWI2 time steps per MODFLOW time step. If NADPTMN is less than 1, it is changed to 1. NADPTMN must be less than or equal to NADPTMX.

ADPTFCT—is the factor used to evaluate tip and toe thicknesses and control the number of SWI2 time steps per MODFLOW time step. When the maximum tip or toe thickness exceeds the product of TOESLOPE or TIPSLOPE the cell size and ADPTFCT, the number of SWI2 time steps are increased to a value less than or equal to NADPT. When the maximum tip or toe thickness is less than the product of TOESLOPE or TIPSLOPE the cell size and ADPTFCT, the number of SWI2 time steps is decreased in the next MODFLOW time step to a value greater than or equal to 1. ADPTFCT must be greater than 0.0 and is reset to 1.0 if NADPTMX is equal to NADPTMN.

NU—Values of the dimensionless density

     ISTRAT = 1 – Density of each zone (NSRF+1 values).
     ISTRAT = 0 – Density along top of layer, each surface, and bottom of layer (NSRF+2 values)

ZETA—Initial elevations of the active surfaces.

SSZ—Effective porosity

ISOURCE—Source type of any external sources or sinks, specified with any outside package (i.e. WEL Package, RCH Package, GHB Package). There are three options:

If ISOURCE > 0 – Sources and sinks have the same fluid density as the zone ISOURCE. If such a zone is not present in the cell, sources and sinks have the same fluid density as the active zone at the top of the aquifer.

If ISOURCE = 0 – Sources and sinks have the same fluid density as the active zone at the top of the aquifer.

If ISOURCE < 0 – Sources have the same fluid density as the zone with a number equal to the absolute value of ISOURCE. Sinks have the same fluid density as the active zone at the top of the aquifer. This option is useful for the modeling of the ocean bottom where infiltrating water is salt, yet exfiltrating water is of the same type as the water at the top of the aquifer.

OBSNAM—is a string of 1 to 12 nonblank characters used to identify the observation. The identifier need not be unique; however, identification of observations in the output files is facilitated if each observation is given a unique OBSNAM.

LAYER—is the layer index of the cell in which the ZETA observation is located.

ROW—is the row index of the cell in which the ZETA observation is located.

COLUMN—is the column index of the cell in which the ZETA observation is located.

Appendix 2.   List of Selected Input Datasets for Example Simulation 1

The NAM, DIS, BAS, LPF, and SWI2 input datasets for example simulation 1 are presented to provide users with a quick reference for setting up a MODFLOW model that uses the SWI2 Package. Some brief annotations have been added as comments within the SWI2 dataset to help the reader understand the purpose of various sections of the input dataset. Comments in the SWI2 dataset are identified with a "#" in column 1. The MODFLOW-2005 documentation (Harbaugh, 2005) provides information about variables and data contained in the other MODFLOW packages. Font sizes of the input datasets have been reduced so that lines will fit within page margins.

File name: `swiex1.nam`

```
# Name file for mf2005, generated by Flopy.
LIST    2 swiex1.list
DIS   11 swiex1.dis
BAS6  13 swiex1.bas
LPF   15 swiex1.lpf
WEL   20 swiex1.wel
SWI2  29 swiex1.swi
DATA(BINARY)   55 swiex1.zta REPLACE
DATA(BINARY)   56 swiex1.swb REPLACE
OC   14 swiex1.oc
DATA(BINARY)   51 swiex1.hds REPLACE
DATA(BINARY)   52 swiex1.ddn REPLACE
DATA(BINARY)   53 swiex1.cbc REPLACE
PCG  27 swiex1.pcg
```

File name: `swiex1.dis`

```
# Discretization file for MODFLOW, generated by Flopy.
         1          1         50          1          4          2
0
         0  5.000e+00              (5G13.0)            -1 DELR(NCOL)
         0  1.000e+00              (5G13.0)            -1 DELC(NROW)
        11          1              (5G13.0)            -1 TOP OF SYSTEM
             0          0          0          0          0
             0          0          0          0          0
             0          0          0          0          0
             0          0          0          0          0
             0          0          0          0          0
             0          0          0          0          0
             0          0          0          0          0
             0          0          0          0          0
             0          0          0          0          0
             0          0          0          0          0
        11          1              (5G13.0)            -1 BOTTOM OF LAYER
           -40        -40        -40        -40        -40
           -40        -40        -40        -40        -40
           -40        -40        -40        -40        -40
           -40        -40        -40        -40        -40
           -40        -40        -40        -40        -40
           -40        -40        -40        -40        -40
           -40        -40        -40        -40        -40
           -40        -40        -40        -40        -40
           -40        -40        -40        -40        -40
           -40        -40        -40        -40        -40
      400.000000          200  1.000000 SS
```

File name: `swiex1.bas`

```
# Basic package file for MODFLOW, generated by Flopy.
FREE
        13          1              (5I4)        -1 IBOUND Array for Layer
    1   1   1   1   1
    1   1   1   1   1
    1   1   1   1   1
    1   1   1   1   1
    1   1   1   1   1
    1   1   1   1   1
    1   1   1   1   1
    1   1   1   1   1
    1   1   1   1   1
    1   1   1   1  -1
-999.990000
        13          1              (5G13.0)     -1 Starting Heads in Layer
        0          0          0          0          0
        0          0          0          0          0
        0          0          0          0          0
        0          0          0          0          0
        0          0          0          0          0
        0          0          0          0          0
        0          0          0          0          0
        0          0          0          0          0
        0          0          0          0          0
        0          0          0          0          0
```

File name: `swiex1.lpf`

```
# LPF for MODFLOW, generated by Flopy.
        53  -1.0e+30           0
0
0
1
0
0
        15          1              (25G3.0)        -1 HK() = Horizontal hydraulic conductivity of layer 1
  2 2 2 2 2 2 2 2 2 2 2 2 2 2 2 2 2 2 2 2 2 2 2 2 2
  2 2 2 2 2 2 2 2 2 2 2 2 2 2 2 2 2 2 2 2 2 2 2 2 2
        15          1              (25G3.0)        -1 VKA() = Vertical hydraulic conductivity of layer 1
  2 2 2 2 2 2 2 2 2 2 2 2 2 2 2 2 2 2 2 2 2 2 2 2 2
  2 2 2 2 2 2 2 2 2 2 2 2 2 2 2 2 2 2 2 2 2 2 2 2 2
```

File name: `swiex1.swi`

```
# Salt Water Intrusion package file for MODFLOW-2005, generated by Flopy.
#--Dataset 1
         1          1          0         55         56          0
#--Dataset 2a
         1          0          3
#--Dataset 3a
 2.000000e-01  2.000000e-01
#--Dataset 4
        29          1              (10G13.0)        -1
             0      0.025
```

```
#--Dataset 5
        29              1           (10G13.0)          -1
            0              0              0              0              0              0              0              0              0
            0              0              0              0              0              0              0           -2.5           -7.5
        -12.5          -17.5          -22.5          -27.5          -32.5          -37.5            -40            -40            -40
          -40            -40            -40            -40            -40            -40            -40            -40            -40
          -40            -40            -40            -40            -40            -40            -40            -40            -40
          -40            -40            -40            -40            -40
#--Dataset 6
 CONSTANT  2.000e-01          (10G13.0)          -1
#--Dataset 7
        29              1           (10I13)          -1
            2              1              1              1              1              1              1              1              1
            1              1              1              1              1              1              1              1              1
            1              1              1              1              1              1              1              1              1
            1              1              1              1              1              1              1              1              1
            1              1              1              1              1              1              1              1              1
            1              1              1              1              1
```

www.ingramcontent.com/pod-product-compliance
Lightning Source LLC
Chambersburg PA
CBHW081618170526
45166CB00009B/3017